What is mathematical logic?

J. N. CROSSLEY
C. J. ASH
C. J. BRICKHILL
J. C. STILLWELL
N. H. WILLIAMS

What is mathematical logic?

OXFORD UNIVERSITY PRESS

London   Oxford   New York

1972

Oxford University Press, Ely House, London W.1

GLASGOW   NEW YORK   TORONTO   MELBOURNE   WELLINGTON
CAPE TOWN   IBADAN   NAIROBI   DAR ES SALAAM   LUSAKA   ADDIS ABABA
DELHI   BOMBAY   CALCUTTA   MADRAS   KARACHI   LAHORE   DACCA
KUALA LUMPUR   SINGAPORE   HONG KONG   TOKYO

PRINTED IN GREAT BRITAIN
BY J. W. ARROWSMITH LTD.
BRISTOL, ENGLAND

Preface

THE lectures on which this book is based were conceived by Chris Brickhill and John Crossley. Our aim was to introduce the very important ideas in modern mathematical logic without the detailed mathematical work which is required of those with a professional interest in logic. The lectures were given at Monash University and the University of Melbourne in the autumn and winter of 1971. Their popularity induced us to produce this book which we hope will give some idea of the exciting aspects of mathematical logic to those who have no mathematical training.

We must add that we very much enjoyed giving these lectures and that the audience response exceeded all our expectations. We thank Associate Professor John McGechie of Monash University and Professor Douglas Gasking of the University of Melbourne who so ably supported us in this venture. We must also thank Dennis Robinson and Terry Boehm for their assistance to Chris Brickhill in the preparation of these notes. Finally we thank Anne-Marie Vandenberg who so masterfully dealt with the typing.

Ayers Rock   J. N. CROSSLEY
August 1971

Contents

Introduction

MATHEMATICAL logic is a living and lively subject. We hope that this will be conveyed by the somewhat unconventional style in which this book is written. Originally lectures were given by four of the authors but they have been so revamped that it seemed inappropriate to specify the authorship in detail.

It is our hope and belief that any reader who later undergoes a full course in logic would be able to fill out all our sketches into full proofs.

The chapters are in many ways independent, though items from Chapters 2 and 3 are used in Chapters 5 and 6, and we suggest that a reader who finds a chapter difficult should turn to another one and return to the first one later. In this way you may find the difficulties have been ameliorated.

1

Historical Survey

THE different areas in logic emerged as a result of difficulties and new discoveries in a complicated history, so this first chapter is going to describe a flow chart (*see* p. 2). I shall skip rather briefly over the different areas, so do not be worried if you find a lot of strange terminology—it will be explained later on in the book. I am going to view the history as two different streams, both of which are very long: one is the history of formal deduction which goes back, of course, to Aristotle and Euclid and others of that era, and the other is the history of mathematical analysis which can be dated back to Archimedes in the same era. These two streams developed separately for a long time—until around 1600–1700, when we have Newton and Leibnitz and their invention of the calculus, which was ultimately to bring mathematics and logic together.

The two streams start to converge in the 19th century, let us say arbitrarily about 1850, when we have logicians such as Boole and Frege attempting to give a final and definitive form to what formal deduction actually was. Now Aristotle had made rather explicit rules of deduction, but he had stated them in natural language. Boole wanted to go further than this and he developed a purely symbolic system. This was extended by Frege, who arrived at the predicate calculus which turned out to be an adequate logical basis for all of today's mathematics. Perhaps I can dwell a little on this, since symbolism became so important from this point onwards. A little description of what symbolism can look like will help.

Purely logical connectives, such as *and*, *or*, *not* are given symbols such as &, $\lor$, $\lnot$; we need symbols (x, y, z and so on) for variables and also symbols P, Q, R for predicates (or properties or relations). Out of these we make formulae such as this: $P(x) \lor Q(x)$, which is read as saying that x has property P or x has property Q, and this can be quantified by expressing 'for all x' by $\forall x$ and 'there exists an x' by $\exists x$. Thus $\forall x\, P(x)$ says every x has property P.

Now any mathematical domain can be translated into this language with a suitable choice of predicate letters: arithmetic, for instance. We have numbers

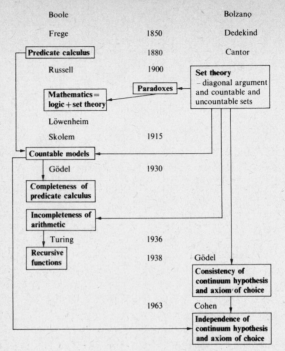

as the objects that the variables will denote, and we have various properties of numbers that we wish to express, such as being equal or the relationship between two numbers and a third one which is the sum of the other two; another one could be the product relationship. You can quickly convince yourself that all statements that we are accustomed to make in number theory about divisibility, prime numbers, and whether one number is the sum of another two can be made using these predicates. Frege gave rules for making deductions in this language and the whole conglomeration is called *predicate calculus*.

Now meanwhile in analysis there was a long period, a couple of centuries, of controversy over the meaning of concepts that Newton introduced—the derivative and the integral—because he talked about infinitesimals. A lot of people did not believe in these and thought they were contradictory, which they were. But nevertheless he got the right results and, to find out why he got the right results, clarification of the notions was made. Some of the people responsible for this were Bolzano, Dedekind, and Cantor. (This brings us down to about 1880.) These people realized that, to deal adequately with derivatives and integrals, infinite sets had to be considered, and considered very precisely. There was no way of avoiding infinite sets. This was the origin of set theory.

I think it is worth pointing out that Cantor had got into set theory from a problem in analysis. He was not trying to define natural numbers or any of the other things that people have used set theory for since. His original motivation was analysis of infinite sets of real numbers. And I think this is really the proper domain of set theory: to solve problems like that rather than problems of definition of primitive concepts. This *can* be done and it was done by Frege (actually in an inconsistent way, but this was put right by Russell). Russell was dedicated to the proposition that mathematics was just logic. Logic to Russell was a lot more than we would consider logic today. We would say that really he showed that mathematics was logic and set theory. With sufficient patience and sufficient lengths of definitions any mathematical field can be defined in terms of logic and set theory and all the proofs carried out within the predicate calculus.

But of course Cantor was jumping ahead at this stage. He went way beyond trying to solve problems in analysis; he was interested in sets themselves and he really discovered how fascinating they were. (His results in set theory did feed back into analysis, as we shall see in a moment.) I think it is important to give you at least two proofs here of Cantor's results because the arguments he used were totally revolutionary and they have permeated the whole of logic ever since then. In fact most of the theorems, I feel, can be traced back to one or other of these arguments that I am about to describe. Cantor, in considering infinite sets, quickly came to the realization that a lot of infinite sets were similar to the set of natural (whole) numbers in the sense that they could be put in one-to-one correspondence with them. This was already known to Galileo in the simple case of even numbers. Galileo realized this correspondence existed and he was rather distraught about it, because he thought this ruined all hope of describing different sizes of infinite sets; he thought the concept must be meaningless. But Cantor was not bothered about this; he said we shall nevertheless say that these two sets are of the same infinite size, and then see how many other sets we can match with the set of natural numbers. And his first major discovery was that the rational numbers can be put in one-to-one correspondence with the natural numbers. (The rational numbers are the fractions p/q, where p and q are natural numbers, $q \neq 0$.)

This comes as a great surprise to people, because as you know the rational numbers lie densely on the line, that is to say, between any two points there is a rational number. So if you try and start counting from the left you run out of numbers before you get anywhere at all. Cantor's method was this: he said that the rational numbers can be arranged as in the table on page 4. In the first row we put all the ones with denominator one, in the second one all the ones with denominator two, and then those with denominator three, and so on. And this table will include all the rational numbers. Now we count them like this: we start at the top left and then zig-zag, following the arrows.

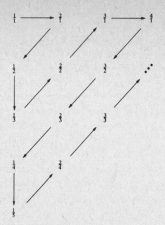

So the list goes $\frac{1}{1}, \frac{2}{1}, \frac{1}{2}, \frac{1}{3}, \frac{2}{2}, \frac{3}{2}, \frac{4}{1}, \frac{3}{2}, \ldots$. We are not going to miss any out by this method, and every rational number will get assigned to a natural number (the number denoting its place in our list). That was his first discovery. In view of this extraordinary fact that a dense set on a line can nevertheless be counted, you might begin to expect that any infinite set can be counted. Of course, as you may well know, this is not the case and this was his second argument.

Now the real numbers, which correspond to the points on the line (all points on the line, forming a continuum) can be expressed by infinite decimal expansions and in general the decimal expansion has to be infinite ($\sqrt{2}$ for instance is an infinite decimal; there is no finite representation for it). So if we are going to make a correspondence between the real numbers and the natural numbers it will look something like this: we start by matching zero with some infinite decimal. (I just consider real numbers between zero and one, so there is nothing in front of the decimal point.) Next one is matched with some infinite decimal, then two is matched with some infinite decimal, and so on. Our ambition is somehow to get a list with all the real numbers in. Cantor said that no matter how you try to do this you fail, for the following reason: for any list that is given we can construct a number that is different from the first number by writing down a different digit in the first decimal place. We can make the number different from the second number by making it different in the second decimal place and we can make it different from the third number by making it different in the third decimal place, and by continuing in this way we get a real number that is an infinite decimal that is different from any number of the list. And it does not matter what the list is, this method will always work. So there cannot be any correspondence between the real numbers and the natural numbers, and so we have discovered a larger infinite set.

Cantor gave an elaboration of this argument too, which I shall briefly describe. Take any set, S: you cannot put S in one-to-one correspondence with the set of all its subsets, that is, the set $\{T: T \subseteq S\}$. And the argument is really the same as the previous one, though it looks a bit different. This time we shall make what we would loosely call a list of the members of S. They might not be a listable set, but just imagine that somehow we are trying to match the members of S with the subsets of S. And let T_s be the subset that we match with the member s. Cantor immediately constructs a subset which is not on this list. He constructs a subset U consisting of the members s which are matched with a subset to which they do not belong. Now if you think about it we are making U different from every subset T because T_s either has s as an element or it does not. Either way we make U different from T_s with respect to this element s. If s is in T_s, we leave s out of U. If s is not in T_s we put s in U. Thus there are more subsets of S than can ever be matched one-to-one with the members of S—any such matching omits a set U such as we have described.

Now there is one further step you can take in this argument: suppose S is the set of all sets in the universe. Take the subsets of S—that is, another set, or it ought to be. This argument seems to be saying that if you start with all of the sets in the universe and take all the subsets, you get more. But you cannot get more than there is in the universe. Cantor was aware of this problem but took it in his stride, which I think was very perceptive of him. Russell also found out about it, and he worried about it. This was the notorious Russell paradox, or something very similar to it anyway.

So this whole program of reducing mathematics to logic and set theory was endangered by the fact that set theory looked very dodgy at this point. Immediately the first concern of mathematicians was to clean up the basis of set theory so that paradoxes could not arise. Somehow we have to prevent the collection of all sets in the universe being considered as a set itself. And that is actually very easy to do, if you insist that sets are always built up from things that you know about to begin with—that is, the natural numbers and the real numbers—and you do not stop; you always consider it possible to go one level higher, so that you never complete the whole universe and only things which are completed are regarded as sets. Now this idea was suggested about 1908, by Zermelo. And Russell did a similar thing, though his method is technically more clumsy and is scarcely used today. The idea was finally polished up with the help of Fraenkel and Skolem in 1922. This was then an axiomatic set theory which seemed to avoid the problems.

Around the same time another problem had developed with this conception of reducing mathematics to a symbolic language, and this was the following. So far, we had a certain domain of study in mind; it might have been the natural numbers or it might have been analysis and we have gone to a formal theory, just using symbols. And so formal that everything is just mechanical.

Every statement about this domain is a series of symbols. We make the deductions by mechanical operations on the symbols. We get things out which we call theorems, and these are true statements about the domain. So that seems fine. But when we just have a collection of symbols, they are susceptible to many different interpretations (or models as they are called) and it is possible that interpretations exist that are entirely different from the domain we thought of in the first place. So to be careful we should consider other models, if there are any, of the theory. The decisive result which shows that there are definitely unintended interpretations was proved in 1915 by Löwenheim, and this was sharpened by Skolem. Naturally enough it is called the Löwenheim–Skolem Theorem.

We shall talk a little about this interesting result. This deserves a box of its own in our chart. The Löwenheim–Skolem Theorem simply says that every theory has countable models. This was a tremendous shock because we already have uncountable sets and theories about these things, such as the theory of real numbers, that are supposed to be about uncountable domains. So we have models for axioms that are definitely not what we intended. I can briefly describe how this result comes about. First we observe that languages after all are only countable—they are just finite strings of symbols and by elaborating the technique of enumerating the rational numbers given above, you can enumerate all the statements in a theory. Now consider the intended domain and what the axioms really determine about it. Axioms can say that certain objects exist initially; for example, they can say that zero exists, if you talk about the natural numbers. If you analyze the other things they can say, roughly speaking they are things like: if such and such exists, then something else must exist. So not too many things are actually necessary in order to satisfy the axioms. We have to start with the initial objects and then what is needed because of them, and what is needed because of that, and so a skeleton can be built up which contains only countably many points. This is possible because, if you can only say countably many things about a certain domain, then only countably many objects are needed to satisfy those statements.

Obvious as this may be, it escaped people's attention for a long time and then it bothered them, of course, that such things can hold about set theory where there are supposed to be uncountably many objects existing. Since the axioms of set theory yield a proof that uncountable sets exist, how can they be satisfied in a countable domain? This state of affairs is called the Skolem paradox. It is not a real paradox and it gets resolved later, but there seems to be a paradox if theories can have countable models and yet there is a theory about uncountable things. We have to say: what does 'countable' mean in relation to a model? And this will be treated in Chapter 3. But this was very thought-provoking and it led to exceptionally good theorems later on.

All the time up until this (1920), and up to 1930 in fact, people had used predicate calculus for making logical deductions without finding out for sure whether they obtained all of the valid statements this way. The argument that was developed for the Löwenheim–Skolem theorem is similar, in some presentations at least, to a proof of the completeness of the predicate calculus. This is another milestone. This was the first major result that Gödel proved, and not the last. What exactly does 'completeness' mean? If we have a language with predicate symbols, variables, and quantifiers, certain things are valid no matter how we interpret the symbols. No matter what property P may denote and no matter what domain the variables are ranging over, $\forall x \, P(x) \rightarrow \exists x \, P(x)$ is definitely true (where $\rightarrow$ means implies). The object of the predicate calculus was to produce in a mechanical fashion *all* logically valid formulae. Gödel succeeded in proving that it was complete, that it did achieve this object.

Just a little while later he proved a really shattering result, in 1931, about incompleteness: incompleteness was present even in arithmetic. By arithmetic I mean the theory I sketched before, in which you can prove things about the natural numbers and the properties that are definable in terms of addition and multiplication. Now, our object with the formal system for arithmetic naturally would be to prove everything that is true about the natural numbers. But Gödel showed that, no matter what formal system you start with, this cannot be accomplished; there are always sentences of arithmetic which cannot be proved or disproved relative to the original system. In establishing this result he used a welter of arguments: some of them dated back to Cantor's diagonal argument; another technique was to express properties of the symbols of the formal system in terms of numbers and hence in terms of statements in the system itself. The system could therefore talk about itself in some sense. And this coding of formulae and symbols was accomplished by rather elaborate machinery: he defined functions from very simple functions like addition and multiplication and from these he built up to more and more complicated functions, but all of these functions could be defined in arithmetic. After he had done this, people began to be interested in the functions themselves. They are called recursive functions and reached their definitive form around 1936. I shall try and briefly explain what a recursive function is. In fact in one word it is a computable function. The functions Gödel used were ones which anybody could compute given simple instructions. Anybody could compute the product of two numbers and the sum of two numbers; anybody can compute the nth prime number. (It is a bit complicated, but you can do it.) Mathematicians began to wonder if Gödel's functions did indeed include every computable function. They were not convinced of this for quite some time, and a lot of experimentation was done with other methods of defining computable.

One of the most important of these is due to Turing; he delineated the concept of a machine, a computing machine. He gave arguments to suggest that any computation a person could do could also be done by one of these machines. He analysed all the steps that you do when you make a computation: writing things down, scanning blocks of symbols, making notes, going back to things you had done before, making lists, and all this sort of thing, and he devised a kind of machine which could do all these operations in a very simple way. The machine only needs to be able to look at squares on a tape and to identify symbols in the squares. It looks at one square at a time and it can move one square to the left or one square to the right and it can change the symbol. It has a program telling it what to do if it sees a certain symbol and it is a finite program; that is all there is to it. (See Chapter 4.) This concept agreed with Gödel's concept which he defined in terms of the language of arithmetic. Others pursued different methods (Church was another to define computable functions) and they all arrived at the same concept. So they finally became convinced that these were, indeed, the computable functions. Now this is an extraordinary thing, because no-one had expected such a vague concept as computability to have a precise definition. And yet it did. This was a great bonanza, because now all sorts of things could be done that could not be done before. For the first time you could prove things to be unsolvable.

What does it mean to have an algorithm or a way of solving a problem? For example, everybody knows the algorithm for solving quadratic equations: the equation $ax^2 + bx + c = 0$ has solutions

$$x = \frac{-b \pm \sqrt{b^2 - 4ac}}{2a}$$

and if you quote the numbers a, b, c, the answer pops up mechanically. So you could get a machine to do this work and it would solve all the infinitely many quadratic equations, all in the same mechanical manner.

There were questions, however, for which no mechanical solution was known. Obviously we do not have a mechanical method of telling whether a given statement of mathematics is true or not. And so this kind of question is open to proof or disproof now for the first time, because we now know what a mechanical method is. It might be possible to prove that there is no mechanical method that would solve such a problem as finding truth or falsity for statements of mathematics. In fact, what about the predicate calculus? We have a way of finding all the valid sentences but it was not known how you could tell if a sentence was valid or not; all that could be said was that if it was valid we should eventually succeed in proving it. And Church showed that there was no mechanical method that would decide truth or falsity in a predicate calculus. This was called the undecidability theorem for predicate calculus. There have been a number of undecidability results since then; of course, all of mathematics, if you take the whole of it, is going to be undecidable. Also a lot of smaller parts, like arithmetic, group theory, and other problems which had been bothering mathematicians for many decades were eventually shown to be undecidable once this precise concept came into existence. And recursive function theory itself is growing in many ways, some of which we shall describe in later chapters (especially Chapter 4).

Now let us get back to set theory. There was another advance in set theory in 1938. Gödel (again) showed the consistency of two particular axioms which I have to talk about a little now. These are the Axiom of Choice and the Continuum Hypothesis. The Axiom of Choice says a very trivial thing. If you try to draw a picture to represent this axiom, it looks utterly trivial. It says: if you have a set of sets then there exists a set consisting of one member from each of these sets. And drawing it in picture form no-one can possibly argue with it, but a lot of things cannot be proved without it. For instance, if you have an infinite set it is a very old theorem that this contains a denumerable subset, and the way you count out an infinite subset is to take one member and since the set was infinite to start off with, there are more members left, so we can take another one; there are still some left, because it is infinite, so we can take a third one, and so on.

We are making an infinite sequence of choices here and it might bother you that there is no way of actually defining this infinite sequence. It bothers me, really, but a lot of mathematicians are not bothered. Here is another example: can you define a function f so that if X is any set of real numbers, then $f(X)$ is a member of X? If you try to do it, you think: first of all, how about taking the least member of X? But that is no good because a lot of sets of real numbers do not have a least member. The set of real numbers greater than zero, for instance, has no least member. So this is not an adequate definition; you can play with more and with more complicated functions like taking the decimal expansions and finding properties of them, and you just never succeed in defining such a function, even though it seems obvious that it must exist. The only way to bring this function into existence is to assert the Axiom of Choice. And since mathematicians frequently need functions like this, they were using the Axiom of Choice for a long time, since before 1900. And that was the first of the axioms that Gödel proved to be consistent. You cannot disprove it, in other words.

The other one was the Continuum Hypothesis. This was a conjecture of Cantor and it is still a conjecture. As soon as Cantor found that the real numbers constituted a bigger set than the natural numbers and he could not seem to find any that were intermediate in size, he thought it would be nice if the continuum was the next biggest infinite set, and made this his conjecture of the Continuum Hypothesis; he expressed it precisely, though.

Gödel showed that this also was consistent.

And the last thing I want to mention (with a big jump in time to 1963) is this: Cohen (*see* Chapter 6) showed that the negations of these two axioms are also consistent. You cannot prove them either. So we are left completely up in the air as to whether they are true or not.

2

The Completeness of Predicate Calculus

IN this chapter I am going to prove the completeness of predicate calculus, that is, that the statements which we can prove in predicate calculus are exactly the ones that are really true. All that I shall assume is that the reader has at least met some of the ideas involved in formalization in a predicate calculus, and so my treatment of this will be correspondingly brief.

Consider the sentence

(i) If x is an ancestor of y and y is an ancestor of z, then x is an ancester of z.

Using '$\rightarrow$' for 'if ... then ...', '&' for 'and' and '$P(x, y)$' for 'x is an ancestor of y' (i) can be formalized within a predicate calculus as

(ii) $P(x, y) \& P(y, z) \rightarrow P(x, z)$.

If we wished to say that everything has an ancestor, we would employ the universal quantifier '$\forall$' to mean 'for all' and the existential quantifier '$\exists$' to mean 'there exists' or 'there is at least one' and write, as a further example of a formalization within a predicate calculus:

(iii) $\forall y \, \exists x P(x, y)$.

With these two examples in mind, I shall now describe predicate calculus formally. To make things more manageable I consider the formal system PC. PC is a predicate calculus with only the one predicate $P(x, y)$. In fact, what I do can be done equally well (it is just more tedious) for systems that have more than the one predicate symbol (e.g. $Q(x, y, z)$ etc.), and which have function symbols and constant symbols as well.

The formal system PC

1. The alphabet of PC: a denumerable set of individual variables: $v_1, v_2, v_3, \ldots$ and a two-place predicate letter P. Two logical connectives: $\rceil$ (not) and & (and). One quantifier symbol $\exists$ (there exists). Three improper symbols: (,) (the left bracket, the comma, and the right bracket).
2. We use these symbols to build the (well-formed) *formulae* of PC, according to the following rules.

(a) If x, y are individual variables, then $P(x, y)$ is a formula of PC.

(b) If ϕ, ψ are formulae of PC then so are $(\phi \And \psi)$ and $\neg\psi$.

(c) If x is an individual variable and ϕ is a formula then so is $\exists x\phi$.

(d) Nothing is a formula of PC unless it is one by virtue of (a), (b), or (c).

For simplicity, we put only the logical symbols $\neg, \And, \exists$ in the alphabet. This is really no restriction, for the other usual symbols, namely $\lor$ (or), $\to$ (implies), $\leftrightarrow$ (if and only if), and $\forall$ (for all) can be defined from these, as follows:

$(\phi \lor \psi)$ is defined as $\neg(\neg\phi \And \neg\psi)$, $(\phi \to \psi)$ as $\neg(\phi \And \neg\psi)$, $(\phi \leftrightarrow \psi)$ as $((\phi \to \psi) \And (\psi \to \phi))$, and $\forall x\phi$ as $\neg\exists x \neg\phi$.

A variable is said to be *bound* if it is governed by a quantifier. If a variable is not bound it is said to be *free*. For example, in the formula $\exists x P(x, y)$, x is bound and y is free.

Bound variables can be changed with impunity if care is taken. That is, we can change variables so long as we do not thereby allow the new variable to be bound by a quantifier that did not bind the old variable. A formula of PC is said to be a *sentence* if it contains no free variables.

Interpretations

Consider the following sentences of PC:

(iv) $\forall x \forall y (P(x, y) \to P(x, y))$,

(v) $((P(x, y) \And P(y, z)) \to P(x, z))$,

(vi) $\forall y \exists x P(x, y)$.

If we interpret P as the ancestor relation over the domain of people, both living and dead, (iv), (v), and (vi) are all true. With this interpretation they become

(iv) if x is an ancestor of y then x is an ancestor of y, for any x and y,

(v) if x is an ancestor of y and y is an ancestor of z then x is an ancestor of z,

(vi) everybody has an ancestor.

Also, if we interpret P as $>$ (greater than) over the natural numbers $(1, 2, 3, \ldots)$ or as $<$ over the integers $(\ldots, -2, -1, 0, 1, 2, \ldots)$ these are all true sentences as well. But if P is interpreted as $<$ over the natural numbers, then (vi) is false. Also, if P is interpreted as 'the father of' over the domain of people, (v) is false. However, (iv) will remain true no matter what interpretation one gives to P. There are many such sentences of PC and they are said to be *universally valid*.

Completeness problem

What one gets if one chooses a finite number of axioms or schemes for axioms (selected formulae) and a finite number of rules of inference (ways of inferring formulae from given finite sets of formulae) in a language such as PC is an example of a formal system. (As we are not giving all the formal details it is not essential for the reader to be concerned with the specific axioms, but at the end of this chapter we list a set of axioms and rule of inference for predicate calculus.)

In what follows, for the sake of definiteness, we shall be thinking of predicate calculus with just one rule of inference (namely *modus ponens*: from ϕ and $(\phi \rightarrow \psi)$ infer ψ). The formulae inferred by means of the rule of inference in a finite number of steps from axioms and previously inferred formulae are called *theorems*. The completeness problem is this: how can we give a finite number of axioms or schemes for axioms from which the rule of inference will give exactly the true sentences?

Before we look at this problem, I shall have to consider formally what is meant by 'truth in an interpretation'. What does it mean to say that a formula is true in an interpretation $\mathscr{A}$? We want to be able to answer this question for all formulae of PC regardless of whether they contain free variables or not (that is, regardless of whether they are or are not sentences). So we give the definition of truth in an interpretation step by step following the earlier definition of formulae of PC. There is one immediate difficulty: $P(x, y)$ when P is interpreted as 'ancestor of' is true for some values we give to x and y and false for others; for example, the ancestor relation holds for some pairs of individuals and not for others. This means that we shall have to specify particular values for free variables when interpreting them. This problem does not arise for bound variables.

So now here is the definition. An interpretation of the formal system PC is a structure $\mathscr{A} = \langle U, R \rangle$, where U is a set (which must not be empty) the members of which are: $a, a_|, a_{||}, \ldots$, and R is a relation on U. U is said to be the universe of the interpretation and we will eventually interpret the predicate letter P as R, so R will itself be two-place.

Now let members of U be assigned to all the individual variables of PC in such a way that not more than one member of U is assigned to each variable of PC. (But we may assign one member of U to two or more variables.) ϕ is said to be satisfied in $\mathscr{A}$ by an assignment of a_1 to x, a_2 to y, ... (where $x, y, \ldots$ are the free variables in ϕ) if the relation over U corresponding to ϕ (that is, when each P in ϕ is replaced by R) holds between the elements assigned to the free variables of ϕ. In this case we write

$$\mathscr{A} \models \phi[a_1, \ldots],$$

where the list in the square brackets includes all the assignments to the free variables in ϕ.

Thus for the interpretation $\mathscr{A} = \langle U, R \rangle$ of PC, we have

1. $\mathscr{A} \models P(x_1, x_2)[a_1, a_2]$ if and only if a_1 bears the relation R to a_2 (which can be written $a_1 R a_2$).
2. $\mathscr{A} \models \neg \phi[a_1, \ldots]$ if and only if it is not the case that $\mathscr{A} \models \phi[a_1, \ldots]$.

3. $\mathscr{A} \models (\phi \And \psi)[a_1, \ldots]$ if and only if $\mathscr{A} \models \phi[a_1, \ldots]$ and $\mathscr{A} \models \psi[a_1, \ldots]$.

4. If $\psi(x_1, \ldots, x_n)$ is of the form $\exists y\phi(x_1, \ldots, x_n, y)$ then $\mathscr{A} \models \psi[a_1, \ldots, a_n]$
   if and only if there is a b in U such that $\mathscr{A} \models \phi[a_1, \ldots, a_n, b]$.

If $\mathscr{A} \models \phi[a_1, \ldots]$ for every possible assignment to the free variables in ϕ then we write $\mathscr{A} \models \phi$. If $\mathscr{A} \models \phi$ then ϕ is said to be *true* in $\mathscr{A}$. If for every interpretation $\mathscr{A}$, $\mathscr{A} \models \phi$ holds, then ϕ is said to be *universally valid*, and we write $\models \phi$.

If ϕ has no free variables, that is, if ϕ is a sentence, we say that $\mathscr{A} \models \phi$ if and only if $\mathscr{A} \models \phi[a_1, \ldots, a_n]$ for some choice of $a_1, \ldots, a_n$. (The choice is in fact irrelevant, as you can easily see.)

As an example of all this, let us take the interpretation $\mathscr{A} = \langle W, A \rangle$ where W is the universe of people, living or dead, and A is the ancestor relation as defined over W. Consider the formula $\exists x P(x, y)$. Let b and c be people. Then

$$\mathscr{A} \models \exists x P(x, y)[b] \quad \text{if and only if} \quad \mathscr{A} \models P(x, y)[c, b]$$
$$\text{for some } c \text{ in } W.$$

This holds if and only if cAb for some c in W. But this is always so: for each b there is always someone who is an ancestor of b, and so there always exists c in W for which cAb. Thus $\mathscr{A} \models \exists x P(x, y)$. That is, $\exists x P(x, y)$ is true in $\mathscr{A}$.

As an exercise, take $\mathscr{N} = \langle N, < \rangle$ where N is the set of natural numbers and show that $\mathscr{N} \not\models \exists x P(x, y)$. (Here $\mathscr{N} \not\models \exists x P(x, y)$ means, of course, it is not the case that $\mathscr{N} \models \exists x P(x, y)$.)

If Σ is a set of sentences of PC and $\mathscr{A}$ is an interpretation such that $\mathscr{A} \models \phi$ for every sentence ϕ in Σ, then $\mathscr{A}$ is said to be a *model* of Σ.

Now our aim is to find a formal system that produces all the universally valid formulae. That is, we have to suggest some formulae as axioms and one or more rules of inference such that the theorems generated by them are all (and, in order to have consistency, only) the universally valid formulae. Once we have the axioms and rules, we formally define a proof in PC to be a finite sequence of formulae such that each one is either an axiom or inferred from formulae earlier in the sequence by a rule of inference. By a *theorem* we mean the last line of any proof and we write $\vdash \phi$ if ϕ is a theorem.

Often we want to add extra axioms to any given system and construct proofs within the larger system. If Σ is the set of axioms, we write $\Sigma \vdash \phi$ if ϕ can be proved in the system that we get by adding the sentences in Σ as axioms to the initial predicate calculus axioms.

We say that a set of sentences Σ is consistent if for no formula ϕ do we have both $\Sigma \vdash \phi$ and $\Sigma \vdash \neg\phi$.

THEOREM: If Σ is consistent then if $\Sigma \not\vdash \neg\phi$ then $\Sigma + \phi$ is consistent. (We write $\Sigma \not\vdash \psi$ for it is not the case that $\Sigma \vdash \psi$, and $\Sigma + \phi$ means Σ with the additional axiom ϕ.)

In other words, if $\Sigma + \phi$ is inconsistent, then $\Sigma \vdash \neg\phi$.

COMPLETENESS THEOREM: ϕ is provable if and only if ϕ is universally valid, in other words $\vdash \phi$ if and only if $\models \phi$.

Now if we take the axioms (such as those listed at the end of this chapter) it turns out that it is straightforward but laborious to check that each of the axioms is universally valid, and that the rule of inference preserves universal validity. From this we are assured that all the theorems are universally valid. Thus we have that if $\vdash \phi$ then $\models \phi$. This also shows that the predicate calculus is consistent, for if $\vdash \phi$ and $\vdash \neg \phi$ then $\models \phi$ and $\models \neg \phi$. Thus both ϕ and $\neg \phi$ would be universally valid, which is impossible.

Now in order to prove completeness we need to show that if $\models \phi$ then $\vdash \phi$ or equivalently that if ϕ is not a theorem, then ϕ is not universally valid. That is, we need to show that if $\not\vdash \phi$ then $\neg \phi$ has a model, because $\neg \phi$ has a model only if ϕ is not universally valid. But, the above theorem if $\not\vdash \phi$ then $\{\neg \phi\}$ is consistent. All we need is to prove the theorem that follows.

The Gödel–Henkin Completeness Theorem

If Σ is a consistent set of formulae, then there exists an interpretation $\mathscr{A}$ such that $\mathscr{A} \models \psi$ for all ψ in Σ.

For if we then take Σ as just the single sentence $\neg \phi$ we shall have shown that if $\neg \phi$ is consistent then $\mathscr{A}$ is a model for $\neg \phi$. This implies that ϕ is false in $\mathscr{A}$ so that ϕ is not universally valid. I shall now briefly sketch a proof of this theorem, starting by setting out the stages in the proof, which I shall then explain in more detail.

1. | Start with a theory Σ |

2. | Add individual constants to the language $b_1, b_2, \ldots$ (these will be called 'witnesses') |

   Check to see if the theory is consistent when these have been added.

3. | List all formulae with v_1 as free variable: $\psi_0(v_1), \psi_1(v_1), \ldots$ |

4. | Add new axioms of the form $\exists v_1 \psi(v_1) \to \psi(b)$ for an appropriate witness. One new axiom for each formula listed at stage 3.

Check for consistency again.

5. | Apply Lindenbaum's Lemma to get an enlarged set of sentences Σ^* such that $\Sigma^* \vdash \phi$ or $\Sigma^* \vdash \neg\phi$ for each sentence ϕ of the language.

6. | Define an interpretation $\mathscr{A}$ for the extended set of sentences Σ^*.

7. | Check to see that $\mathscr{A} \models \phi$ if and only if $\Sigma^* \vdash \phi$.

8. | Σ is contained in Σ^* so $\mathscr{A} \models \phi$ for all ϕ in Σ, and so $\mathscr{A}$ is a model of the kind we need.

Before I go on, I should explain Lindenbaum's Lemma. A consistent set of sentences Σ is said to be *full* if for each sentence ϕ of the language either $\Sigma \vdash \phi$ or $\Sigma \vdash \neg\phi$. Then Lindenbaum's Lemma states:

If Σ is consistent, then there exists a full consistent extension Σ^* of Σ, that is, for any ϕ in the language either $\Sigma^* \vdash \phi$ or $\Sigma^* \vdash \neg\phi$, but not both.

The proof goes like this: list all the sentences $\phi_1, \phi_2, \ldots$ of PC. We shall build up step by step to Σ^*. Define a sequence $\Sigma_0, \Sigma_1, \Sigma_2, \ldots$ of sets of sentences in the following way. Put $\Sigma_0 = \Sigma$. Take

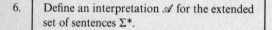

$$\Sigma_1 = \begin{cases} \Sigma_0 & \text{if } \Sigma_0 \vdash \neg\phi_1, \\ \Sigma_0 + \phi_1 & \text{if } \Sigma_0 \nvdash \neg\phi_1. \end{cases}$$

(Thus, if ϕ_1 can be added to Σ_0 so that the new set is still consistent, then add it to get Σ_1; if not, leave $\Sigma_1 = \Sigma_0$.) And in fact do this successively with

$\phi_1, \phi_2, \phi_3, \ldots$ So

$$\Sigma_{n+1} = \begin{cases} \Sigma_n & \text{if } \Sigma_n \vdash \neg\phi_{n+1}, \\ \Sigma_n + \phi_{n+1} & \text{if } \Sigma_n \nvdash \neg\phi_{n+1}. \end{cases}$$

Now each Σ_n is consistent, because we started with a consistent set and were careful to preserve consistency at each addition.

Let Σ^* be what we get after we have added, or not added as the case may be, each formula in our list of sentences of PC. Σ^* will be consistent because, as all proofs are finite in length, a proof of inconsistency in Σ^* will mean that one of the Σ_n is inconsistent. But this is impossible because we have guaranteed by the construction that each of the Σ_n is consistent.

Also all sentences of the language are in the list $\phi_1, \phi_2, \ldots$ and at each stage n we determined whether or not to add ϕ_n to the Σ_n. Only when $\neg\phi_n$ was already provable from this set did we not add it. So for all formulae ϕ either ϕ or $\neg\phi$ can be proved from Σ^* and, in fact, either ϕ or $\neg\phi$ is in Σ^*. Thus Σ^* is a full consistent set, as was asserted to exist in Lindenbaum's Lemma.

Now we can fill in some of the details in the stages of the main proof:

STAGE 2a : Add the constants $b_1, b_2, \ldots$ to the language and revise the formal specification of the language; in other words, revise the alphabet and the definition of a formula of PC. Call what we now have Σ_+. These new constants are added so that, whenever we have a property $\psi(v_1)$ which we know is enjoyed by some object in our universe, we shall fix on some constant b and assert $\psi(b)$. Thus b will be a definite *witness* that there is some element with the property ψ.

STAGE 2b : Because all we are doing is adding names of the objects in a prospective universe, Σ_+ will be consistent if Σ is consistent.

STAGE 3 : List all the formulae with just v_1 as free variable:

$$\psi_1(v_1), \ldots, \psi_n(v_1), \ldots.$$

Let θ_n be the formula

$$\exists v_1 \psi_n(v_1) \rightarrow \psi_n(b),$$

where b is the first witness not previously used in a ψ or a θ.

STAGE 4a : Now we want to add the θ_n as axioms, so let

$$\Sigma^0 = \Sigma_+,$$

$$\Sigma^{n+1} = \Sigma^n + \theta_n,$$

$$\Sigma^\infty = \cup \Sigma^n.$$

Thus Σ^∞ is the system that we get when we add all of the new axioms to Σ.

From the usual axioms it is not difficult to check that each Σ_n is consistent. The essential point is that since b is a new witness, b behaves like a free variable.

STAGE 4b: As each Σ^n is consistent Σ^∞ is consistent, because a proof of Σ^∞'s inconsistency would be a proof of the inconsistency of Σ^n for some n. (All proofs are finite in length and so involve only a finite number of formulae. These finitely many formulae must all lie in some Σ^n.)

STAGE 5a: Applying Lindenbaum's Lemma, we can extend Σ^∞ to a consistent full extension Σ^*.

STAGE 5b: Now for any sentences ϕ and ψ of Σ^* the following are true:
(1) $\Sigma^* \vdash \phi$ or $\Sigma^* \vdash \neg\phi$ (Σ^* is full),
(2) $\Sigma^* \vdash \neg\phi$ if and only if $\Sigma^* \nvdash \phi$ because of (1), since Σ^* is consistent,
(3) $\Sigma^* \vdash (\phi\ \&\ \psi)$ if and only if $\Sigma^* \vdash \phi$ and $\Sigma^* \vdash \psi$,
(4) $\Sigma^* \vdash \exists v_1 \psi(v_1)$ if and only if $\Sigma^* \vdash \psi(b)$ for some b, since $\exists v_1 \psi(v_1) \rightarrow \psi(b)$ is one of the θ_n axioms.

STAGE 6: We now define a model $\mathscr{A} = \langle U, R \rangle$ for Σ^* as follows. The universe $U = \{b_1, b_2, \ldots\}$. The relation R on U is defined by

$$b_i R b_j \quad \text{if and only if} \quad \Sigma^* \vdash P(b_i, b_j).$$

STAGE 7: The steps (1), (2), (3), and (4) parallel 1, 2, 3, and 4 of our inductive definition of truth in an interpretation (on p. 13), so we can conclude that

$$\mathscr{A} \models \phi \quad \text{if and only if } \Sigma^* \vdash \phi.$$

STAGE 8: As Σ is contained in Σ^*, $\mathscr{A} \models \phi$ for all ϕ in Σ.

So we conclude that if Σ is consistent, then Σ has a model. This completes the proof.

We can now establish the Completeness Theorem, stated on p. 15. For if $\nvdash \phi$, then $\{\neg\phi\}$ is consistent, and therefore $\neg\phi$ has a model. In this model ϕ is false, so ϕ is not universally valid. Hence $\models \phi$ implies $\vdash \phi$. This completes the proof.

We finally remark that we have in fact shown more than we set out to do. We have shown that Σ has a *countable model*, since the b_i form a countable set. More will be made of this in the next chapter.

Appendix

Axioms and rule of inference for predicate calculus

We give a finite number of schemes for axioms. (We could make do with a finite number of axioms by using a substitution rule, but in order to simplify the contents of this chapter we have chosen not to.)

The axioms are all formulae of the following forms, where ϕ, ψ, χ are formulae, $x, y, y_1, \ldots, y_n, \ldots$ are variables, and $\phi(y)$ is the result of replacing all free occurrences of x in $\phi(x)$ by y.

$\forall y_1 \ldots \forall y_n(\phi \to (\psi \to \phi))$.

$\forall y_1 \ldots \forall y_n((\phi \to (\psi \to \chi)) \to ((\phi \to \psi) \to (\phi \to \chi)))$.

$\forall y_1 \ldots \forall y_n((\neg \phi \to \neg \psi) \to ((\neg \phi \to \psi) \to \phi)))$.

$\forall y_1 \ldots \forall y_n(\forall x(\phi \to \psi) \to (\phi \to \forall x\psi))$ provided ϕ has no free occurrence of x.

$\forall y_1 \ldots \forall y_n((\phi \to \psi) \to (\forall y_1 \ldots \forall y_n\phi \to \forall y_1 \ldots \forall y_n\psi))$.

$\forall y_1 \ldots \forall y_n(\forall x\phi(x) \to \phi(y))$, provided that when y is substituted for free occurrences of x in $\phi(x)$, the y's are free in $\phi(y)$ (that is, they are not governed by y quantifiers already occurring in ϕ).

The rule of inference is *modus ponens*: from $\phi, (\phi \to \psi)$ infer ψ.

3
Model Theory

IN this chapter I am going to discuss Model Theory, and in doing so I shall be concerned with three separate things. Firstly I want to discuss Predicate Calculus with Identity, which I shall abbreviate as PC(=). Secondly I shall consider the Compactness Theorem and thirdly I shall discuss the Löwen-heim–Skolem Theorems. Model Theory is the study of the relations between languages and the world, or more precisely between formal languages and the interpretations of formal languages. I assume you know what formal languages are but I shall talk a little more about interpretations before I start on PC(=).

We shall begin by looking at an English sentence and try to find what we need in the way of predicate symbols, names, connectives, and quantifiers in order to formalize it. Consider the sentence:

(i) If the boss is in charge and Joe is the boss, then Joe is in charge.

To formalize (i) I am going to take the predicate language $\mathscr{L}$. $\mathscr{L}$ has, in addition to the usual sort of machinery of propositional calculus plus quanti-fiers, a one-place predicate letter P to be interpreted as 'is in charge', a two-place predicate letter E, interpreted as 'is the same as' and two constant letters j and b, for 'Joe' and 'the boss' respectively. We shall also need in this case just two of the usual propositional calculus connectives, & and $\rightarrow$ for 'and' and 'implies'; then (i) can be formalized as

(ii) $$P(b) \,\&\, E(j, b) \rightarrow P(j).$$

The predicate letter E is called the identity predicate and predicate calculus with E is called predicate calculus with identity. So (ii) is a sentence in PC(=).

Now consider the structure $\mathscr{A} = \langle A, R, a_1, a_2, S \rangle$ where A is a non-empty set, R is a property, S is a relation on A, and a_1 and a_2 are members of A. Remember that a property on a set is just a subset of that set and a relation on a set is a set of ordered pairs whose members come from that set. We are later going to say that S is the identity relation on A, so S will be the set of ordered pairs such that the first member of an ordered pair is the same as that pair's second member, and each member of A will be a member of some pair in S; that is, $S = \{\langle x, x \rangle : x \in A\}$. Now $\mathscr{A}$ is a structure in which it is possible to interpret (ii). That means we can let j be interpreted as a_1, b as a_2, E as S, and P as R (for the details, see Chapter 2). We now consider this interpretation and for simplicity we shall take A to be the set with just a_1 and a_2 as members. Then if S is the identity relation on A, what (ii) says is that if a_2 has the property R and if a_1 and a_2 are *the same* then a_1 has the property R; which is clearly true in $\mathscr{A}$. Now it is not necessary to interpret the two-place predicate letter, E, as the identity relation, but if we do, and this is always possible, we say that the interpretation is normal. Thus a *normal* interpretation for a language $\mathscr{L}$ is one for which the interpretation of the two-place predicate, E, is the identity relation. Clearly (ii) is true in every normal interpretation, but not necessarily true in other interpretations. The following are also true in every normal interpretation:

(iii) $\forall x\, E(x, x)$,
(iv) $\forall x\, \forall y(E(x, y) \rightarrow E(y, x))$,
(v) $\forall x\, \forall y\, \forall z(E(x, y)\ \&\ E(y, z) \rightarrow E(x, z))$,
(vi) For any formula ϕ of $\mathscr{L}$, $\forall x\, \forall y(E(x, y) \rightarrow (\phi(x, x) \rightarrow \phi(x, y)))$.

(iii) says that E is reflexive, (iv) that it is symmetric, (v) that it is transitive and (vi) is Leibnitz' Law. (iii), (iv), (v) and (vi) are called the *axioms for equality*.

I now claim that a formula ϕ of $\mathscr{L}$ is true in every normal interpretation if and only if ϕ is provable from the logical axioms together with the axioms for equality. Now writing {Logical} and {Equality} for the sets of the logical and equality axioms respectively, we claim that

{Logical} + {Equality} $\vdash \phi$ if and only if $\mathscr{A} \vDash \phi$ for all normal models $\mathscr{A}$, where ϕ is a formula of $\mathscr{L}$. This is just the claim that $\mathscr{L}$ together with its axioms is complete with respect to its normal interpretations.

Now because the sentences of {Logical} + {Equality} are true in all normal interpretations of $\mathscr{L}$, and because the rules of inference yield, for any interpretation, only true formulae from true formulae, we know that if {Logical} + {Equality} $\vdash \phi$, then $\mathscr{A} \vDash \phi$ for all normal interpretations $\mathscr{A}$. Now we want to show that if ϕ is true in all normal models, then {Logical} + {Equality} $\vdash \phi$. We shall use the Gödel–Henkin completeness theorem in proving this. So suppose {Logical} + {Equality} $\nvdash \phi$. Then {Logical} + {Equality} + $\{\neg \phi\}$ is consistent and so by the Gödel–Henkin theorem this set of sentences will have a model. In order to prove my claim I have to give a normal model for $\neg \phi$. This will show that ϕ is not true in all normal models, thus showing that if ϕ is true in all normal models, then ϕ is provable from {Logical} + {Equality}.

I am now going to outline how one can always get a normal model for a set of sentences including {Equality} given that the set has a model at all. Consider a set A with members $x, y, z, \ldots$ and let there be a relation S defined on A which is reflexive, symmetric, and transitive and such that (vi) holds. (This relation is not necessarily equality, for consider the relation 'same age as' and the domain of people as the interpretation for a language where there are no other predicate symbols. That two people are the same age does not imply that they are the same person. Yet this relation is reflexive, symmetric, and transitive and Leibnitz' Law holds of it, since the interpretation of a formula involving only the symbol E can clearly depend only on the ages of the people involved.) Now divide A up into subsets according to whether the elements are in the relation S to each other or not: x and y belong to the same subset if and only if xSy. We can see that any element bears the relation S to every other element in its subset and to no other from the fact that the axioms for equality hold. Now let us choose a representative from each subset. These representatives will form a normal model for the original set of sentences. All the sentences will be true of them and as the representatives are themselves a subset of the original model, for any two elements in the new model u and v, uSv if and only if $u = v$. So E can be interpreted as identity in this new model. This shows that for any set of sentences, in a language with the two-place predicate letter E, and which contains the axioms for equality, if the set has a model, then we can always find a normal model for it. Thus because {Logical} + {Equality} + $\{\neg\phi\}$ has a model, it has a normal model. So ϕ is not true in every normal model. Thus if $\mathscr{A} \models \phi$ for all normal models $\mathscr{A}$, then {Logical} + {Equality} $\vdash \phi$. This proves my claim.

In Model Theory we are usually interested in particular interpretations. One that I am going to give now is of particular importance as you will soon see. I want to look at the structure $\mathscr{N} = \langle N, < \rangle$ where N is the set of natural numbers, $0, 1, 2, 3, 4, \ldots$ etc. and $<$ is the relation 'less than'. (I should write $\langle N, <, = \rangle$ but in all cases from now on I assume that each interpretation has the identity relation to interpret the equality symbol. I shall also mean, when talking about provability and consistency, provability from and consistency with {Logical} + {Equality}.)

Which sentences of the language $\mathscr{L}$ that we talked about before are true in this interpretation? Some are:

(viii) $\forall x\, (\neg P(x, x))$,

(ix) $\forall x\, \forall y\, (\neg(P(x, y)\, \&\, P(y, x)))$,

(x) $\forall x\, \forall y\, \forall z\, (P(x, y)\, \&\, P(y, z) \to P(x, z))$,

(xi) $\forall x\, \forall y\, (P(x, y)\, \lor\, P(y, x)\, \lor\, E(x, y))$,

(xii) $\exists x\, \forall y\, (\neg P(y, x))$,

(xiii) $\forall x\, \exists y\, (P(x, y)\, \&\, \forall z\, (\neg P(x, z)\, \&\, P(z, y)))$,

(xiv) $\forall x\, (\exists y\, P(y, x) \to \exists y\, (P(y, x)\, \&\, \forall z\, (\neg(P(x, z)\, \&\, P(z, y)))))$.

I am going to refer to these sentences often from now on, so I shall use Σ to refer to the collection of them: Σ is the set $\{(viii), \ldots, (xiv)\}$.

Now (viii), (ix), (x), (xi) are just the things that are true in all ordered sets. That is if P is a relation that orders a set of objects then P is irreflexive (viii), P is asymmetric (ix), P is transitive (x), and if any two objects are not equal then they are comparable and the relation P holds one way between them (xi). (xii) says that our interpretation has a first element and (xiii) says that for each object there is always another that is next biggest in terms of our ordering relation. That is, each element has a successor. Finally (xiv) says that every element except the first has a next smallest. That is, all elements but the first have predecessors. All these sentences are true in $\mathcal{N} = \langle N, < \rangle$ as you can see by looking at them and considering what they say.

I now state something about the set of sentences Σ which will bring out the importance of Σ for $\mathcal{N}$: Σ axiomatizes $\mathcal{N}$ completely. More precisely,

LEMMA. If $\mathcal{A}$ is any model for Σ then the same sentences are true in $\mathcal{A}$ as are true in $\mathcal{N}$.

This lemma can be proved by some rather technical model theory, and from it we can deduce the corollary that $\mathcal{N} \models \psi$ if and only if $\Sigma \vdash \psi$. For suppose $\Sigma \nvdash \psi$. Then $\Sigma + \{\neg \psi\}$ is consistent. Therefore by the Gödel–Henkin theorem it has a model, say $\mathcal{A}$. Thus $\mathcal{A} \models \neg \psi$, and by the lemma we conclude $\mathcal{N} \models \neg \psi$. Thus if $\mathcal{N} \models \psi$ we conclude $\Sigma \vdash \psi$. We had above that the sentences in Σ are all true in $\mathcal{N}$ and as the only rule of inference we are using preserves truth in an interpretation, we also have that if $\Sigma \vdash \psi$ then $\mathcal{N} \models \psi$. Because of the fact that if (and only if) a sentence is true in $\mathcal{N}$ we can prove it from the set of sentences Σ, what we have in fact done is axiomatize the set of sentences true in $\mathcal{N}$. That is, all sentences that are true in $\mathcal{N}$ can be proved from the logical axioms and the axioms of equality plus Σ.

I am going to ask the question: to what extent does Σ determine $\mathcal{N}$ itself? Another way to put this question is to ask: given that we know the sentences of Σ are satisfied in an interpretation, how close are we to knowing that the interpretation is $\mathcal{N}$? First I want to say that $\mathcal{N}$ is not the only model for Σ. Remember that $\mathcal{N}$ is the structure $\langle \{0, 1, 2, 3, \ldots\}, < \rangle$. It is easy to get another structure that is isomorphic to $\mathcal{N}$, that is, has the same structure as $\mathcal{N}$, and such that it will be a model for Σ. All we have to do is guarantee that we have a first element, that every element has a successor, and all but the first has a predecessor, etc., and that the relation of the structure is irreflexive, asymmetric, and transitive and that the members of the set are connected (xi above). Now there was no reason for starting off with the natural number 0 when giving $\mathcal{N}$. So let us consider a new structure $\mathcal{M} = \langle \{1, 2, 3, \ldots\}, < \rangle$ and put $M = \{1, 2, 3, \ldots\}$. $\mathcal{M}$ is clearly isomorphic to $\mathcal{N}$ under the correspondence

$$\{0, 1, 2, 3, \ldots\}.$$

$$\downarrow \downarrow \downarrow \downarrow$$

$$\{1, 2, 3, 4, \ldots\}.$$

That is, we make any member x of the set N in $\mathcal{N}$ correspond to the member $x + 1$ of the set M in $\mathcal{M}$ and in doing so use up all the members of each set. Thus $\mathcal{N}$ and $\mathcal{M}$ are isomorphic models for Σ. Now if we could prove that all the models for Σ were isomorphic to $\mathcal{N}$, then we would have achieved a great deal and we could say that in many important respects Σ did determine $\mathcal{N}$. This is so because isomorphic models differ only in the nature of their elements—the structure in each is the same. That is, the properties and relations that hold in one of two isomorphic models correspond to the properties and relations that hold between the corresponding elements of the other. The difference between two isomorphic models is only in a trivial sense a mathematical or logical one.

But alas, there is also a model for Σ that is not isomorphic to $\mathcal{N}$. To see this, firstly draw the members of the set N out along a straight line:

$$\begin{array}{ccccc} + & + & + & + & + \\ 0 & 1 & 2 & 3 & 4 \;\ldots \end{array}$$

Now between 0 and 1 we mark the points $0, \frac{1}{2}, \frac{2}{3}, \frac{3}{4}, \ldots$, that is, we mark in the points that correspond to the members of the set $\{1 - 1/n : n \in N, n > 0\}$. Next between 1 and 2 we mark in the points corresponding to the fractions $1\frac{1}{2}, 1\frac{1}{3}, 1\frac{1}{4}, 1\frac{1}{5}, \ldots$. That is, we mark in the members of the set $\{1 + 1/n : n \in N, n > 1\}$. Then between 2 and 3 we do what we did for 0 and 1. We mark in the points corresponding to the members of the set $\{3 - 1/n : n \in N, n > 0\}$. So we get a line that looks like:

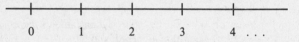

$$0 \qquad \tfrac{1}{2} \;\; \tfrac{2}{3}\tfrac{3}{4}\ldots 1 \ldots 1\tfrac{1}{3} \qquad 1\tfrac{1}{2} \qquad 2 \qquad 2\tfrac{1}{2} \qquad 2\tfrac{2}{3}\ldots 3$$

We then consider the set which has as its members all these points between 0 and 3, including 0 but excluding 1 and 3. This will be the set

$$B = \{1 - 1/n : n \in N, n > 0\} \cup \{1 + 1/n : n \in N, n > 1\} \cup \{3 - 1/n : n \in N, n > 0\}.$$

Now I claim that $\mathcal{B} = \langle B, < \rangle$ is a model for Σ, and that it is not isomorphic to $\mathcal{N}$.

Why is $\mathcal{B}$ a model for Σ? Each of the sentences of Σ is true in $\mathcal{B}$ as you can see by inspection. None of these numbers is smaller than itself, so $\forall x\, (\neg x < x)$ is true. No number is both smaller than and greater than another, so (ix) is true. Also $<$ is transitive over B and any two elements of B are related, so (x) and (xi) are true in $\mathcal{B}$. 0 is a first element, so (xii) is true and each element has a successor and all but zero has an immediate predecessor, so (xiii) and (xiv) are true. Thus all of the sentences in Σ are true and $\mathcal{B}$ is therefore a model for Σ. But clearly $\mathcal{B}$ and $\mathcal{N}$ possess different structures. There can be no element of N corresponding to $1\frac{1}{2}$ in the structure $\mathcal{B}$, since this has infinitely many elements before it and no element of N has this property. So the structures cannot be isomorphic. We conclude that not only does Σ have $\mathcal{N}$ as a model, it has other models isomorphic to $\mathcal{N}$ and yet other models not isomorphic to $\mathcal{N}$. So from the last result we know that Σ cannot tell us very much about $\mathcal{N}$. If all we are told is that we have a structure and that the sentences in Σ are true in this structure, then we cannot say whether the elements in it are the natural numbers, or even behave like the natural numbers in *all* respects. Σ neither determines what objects we have nor encompasses all the structural relations into which they enter. The only thing we do know is that we can prove from Σ plus the logical and equality axioms any sentence of our formal language which is true in $\mathcal{N}$.

I want to stop here for the moment and turn to another thing. I am now going to tell you what the Compactness Theorem is and what it says. But before I do, I shall have to prove a lemma.

LEMMA. If Σ is inconsistent, then some finite subset of Σ is inconsistent.

If Σ is inconsistent, then by definition we have that for some formula ϕ both $\Sigma \vdash \phi$ and $\Sigma \vdash \neg\phi$, or in other words, $\Sigma \vdash \phi \,\&\, \neg\phi$. That is, there is a finite string of formulae $\phi_1, \phi_2, \ldots, \phi_n$ where $\phi_n = \phi \,\&\, \neg\phi$ and where for each ϕ_i either

(i) ϕ_i is a logical or equality axiom, or
(ii) ϕ_i is a member of Σ, or
(iii) ϕ_i follows by the rule of inference from two formulae in the list before ϕ_i.

Because the list is finite, the number of formulae in it that are members of Σ will be finite. So we conclude that if Σ is inconsistent, then there is a finite subset of Σ that is inconsistent.

Now from the Gödel–Henkin theorem that was proved in the last chapter, we know that if Σ is consistent (and Σ may be infinite) then Σ has a model. So now I want to prove, using the lemma I just proved and the Gödel–Henkin theorem, the Compactness Theorem.

THE COMPACTNESS THEOREM. If every finite subset of Σ has a model then Σ has a model.

Assume that every finite subset of a possibly infinite set of sentences Σ has a model. Then we know that every finite subset of Σ is consistent. But then by the lemma Σ must be consistent. So by the Gödel–Henkin theorem Σ has a model.

As an example of this I am going to show you a set of sentences that is infinite and which is such that every finite subset has a model. Thus by the Compactness Theorem we shall deduce that the set itself has a model.

Think of the language $\mathcal{L}$ that I discussed earlier. It was the predicate calculus with identity with one additional predicate letter. I said that the structure $\langle N, < \rangle$ was an interpretation for $\mathcal{L}$. I now want to consider an extension of the language $\mathscr{L}$, I shall call it $\mathscr{L}^+$. $\mathscr{L}^+$ is to be the language $\mathscr{L}$ with an additional constant letter c. Also I want to consider a particular collection of sentences of $\mathscr{L}^+$ in order to see what this collection of sentences says about the models of $\mathscr{L}^+$. The sentences are those in Σ together with

ψ_1 $\exists v_1 \, P(v_1, c)$

ψ_2 $\exists v_1 v_2 (P(v_1, v_2) \,\&\, P(v_2, c))$

ψ_3 $\exists v_1 v_2 v_3 (P(v_1, v_2) \,\&\, P(v_2, v_3) \,\&\, P(v_3, c))$

$\vdots$

ψ_n $\exists v_1 v_2 \ldots v_n (P(v_1, v_2) \,\&\, \ldots \,\&\, P(v_{n-1}, v_n) \,\&\, P(v_n, c))$

$\vdots$

Now suppose that Σ^* is the collection of all those sentences. That is, Σ^* is the set of sentences in Σ together with all the sentences ψ_n for $n \geqslant 1$. Suppose that Σ' is a finite subset of Σ. I am going to consider a collection of structures $\langle A, R, a \rangle$ where a is an element of A and I am going to show how some of these can be models for Σ'. If I produce a model for each Σ' I shall have produced a model for each finite subset of Σ^* and thus, using the compactness theorem, shall have shown that Σ^* itself has a model.

I first stipulate that the A and the R of $\langle A, R, a \rangle$ are to be N and $<$ of my structure $\mathscr{N} = \langle N, < \rangle$. Thus $\langle A, R \rangle$ is a model for Σ. Secondly, consider a finite subset Σ' of Σ^*. Σ' will involve some of the sentences in Σ and only finitely many of the ψ's. Let k be the largest n such that $\psi_n \in \Sigma'$. Now I claim that $\langle N, <, k \rangle$ is a model for Σ'. First all the sentences in Σ are true in $\langle N, <, k \rangle$ because they are true in $\langle N, < \rangle$. So the sentences in Σ' that are also in Σ are true in $\langle N, <, k \rangle$. Secondly we have that if $n \leqslant k$ then $\langle N, <, k \rangle \vDash \psi_n$. This is clearly so because, for instance, ψ_1 says that there is something that is less than c. And if we interpret c as 1, ψ_1 is true. If $k = 1$ we can interpret c as 1, so the interpretation where $k = 1$ is a model for ψ_1. ψ_2 says that there are two things such that the first is less than the second and the second is less than c. Thus ψ_2 is true in $\langle N, <, 2 \rangle$, and so on. Each finite subset will have a largest ψ_n, so each finite subset will have models $\langle N, <, k \rangle$ where k is at least as big as any n of a ψ_n in the finite subset. So we have that Σ^* has a model by the Compactness Theorem.

In this model, let's call it $\mathscr{A}$, the same sentences in the language $\mathscr{L}$ will be true as were true in $\mathscr{N}$, because we had at the beginning of this chapter that $\langle N, < \rangle \vDash \psi$ if and only if $\Sigma \vdash \psi$. But now let us consider what the model for Σ^* looks like. $\mathscr{A}$ will have a first, second, third element and so on because of the axioms in Σ. For simplicity let us just call these $0, 1, 2, \ldots$. Now c of $\mathscr{L}$ must be assigned to an object of the domain of $\mathscr{A}$. But this cannot be a member of N, because for any n, if $c = n$ where $n \in N$, then ψ_{n+1}

will say that $c > n$ which is impossible. Thus the model for Σ^* cannot contain only natural numbers, but must also have something bigger than all the natural numbers (to give the interpretation of c). Here we have what is called a *non-standard* model for Σ, so called because we can easily show that it is not isomorphic to the intended, or *standard* model $\mathcal{N}$ of Σ.

In fact we have already described such a non-standard model, but this technique can be used more generally. For example, it is not easy to give a full description of a non-standard model for the sentences true in $\langle N, <, +, \cdot \rangle$, but the method we have just used can be applied to show that one exists. In a similar fashion, the Compactness Theorem can be used to yield non-standard models for other number systems. In the case of the real numbers such a model can be used to show, for example, the consistency of the use of infinitesimals.

To get the flavour of the compactness theorem, let us pause for another example of its use. Let Σ be a set of sentences with arbitrarily large finite normal models, then Σ has an infinite normal model. To prove this, consider, in the language of Σ augmented by an *infinite* set of new constant letters $c_1, c_2, \ldots$, the set of sentences Σ^* consisting of Σ together with all those sentences $\neg E(c_i, c_j)$ for which $i \neq j$. We show that Σ^* has a model by showing that every finite subset of Σ^* has, and then applying the compactness theorem. For suppose that Σ' is a finite subset of Σ^*. Then Σ' contains, apart from some of the sentences of Σ, finitely many sentences $\neg E(c_i, c_j)$. These will involve only finitely many of the constant letters c_i, which will, for some n, be among $c_1, \ldots, c_n$. Now, by assumption, Σ has a normal model $\langle A, \ldots \rangle$ with at least n elements, so choosing elements $a_1, a_2, \ldots$ of A with $a_1, \ldots, a_n$ distinct, it is easy to see that $\langle A, \ldots, a_1, a_2, \ldots \rangle$ is a model for Σ', where $a_1, a_2, \ldots$ are the interpretations of $c_1, c_2, \ldots$. Thus Σ^* has a model, and so has a normal model, $\langle B, \ldots, b_1, b_2, \ldots \rangle$, where $b_1, b_2, \ldots$ are the interpretations of $c_1, c_2, \ldots$. Then $\langle B, \ldots \rangle$ is a normal model of Σ, since Σ^* includes Σ, and $b_i \neq b_j$ whenever $i \neq j$, by the remaining sentences of Σ^*, so B is infinite. Thus Σ has an infinite normal model as required.

The Compactness Theorem is, for the reasons I have indicated and many others, an important tool in model theory. I shall shortly turn to another such tool, but before this I must introduce some notions in set theory. I want to talk about comparing the sizes of infinite sets. There are many ways of comparing infinite sets. One way is to look and see whether one set is included in the other. For example, we know that $\{0, 1, 2, \ldots\}$ is a subset of $\{-1, 0, 1, 2, \ldots\}$ and we write $\{0, 1, 2, \ldots\} \subseteq \{-1, 0, 1, 2, \ldots\}$. Another way is to see if there is a one-to-one correspondence between the members of the two sets. That is, we see if we can pair off the members of the two sets so that each member from one set is paired with just one member from the other and there are no members in either set left unpaired. For example, the above two sets can be paired off: 0 from the first with -1 of the second, 1 from the

first with 0 from the second, 2 with 1, then 3 with 2, and so on—in general
n with $n - 1$. In this way all members of one set are paired off, each with
just one member from the other and there are no unpaired members.

For our purposes it is better if we say that the sets are the same size if there
is a way of pairing as I have just described. So for our purposes, the sets
$\{-1, 0, 1, 2, 3, \ldots\}$ and $\{0, 1, 2, 3, \ldots\}$ are the same size. Two sets which are
the same size in this sense are said to *have the same cardinal* (*see also* Chapter 6).

A set with the same cardinal as the set of natural numbers is said to be
denumerable, so the set $\{-1, 0, 1, 2, 3, \ldots\}$ is denumerable. A set which is
either denumerable or finite is said to be countable. Now it was shown in
Chapter 1, that the set of real numbers is not countable. (I shall not repeat
the proof here.) So there is at least one uncountable size of infinite set, and
in fact there are many.

The proof of the Gödel–Henkin theorem showed that not only did every
consistent set of sentences in a countable language (that is, a language with
a countable number of formulae) have a model, it also showed that it had a
countable model. Although this model was not necessarily normal, by an
argument similar to the one presented earlier in the chapter, we know that
we can always find a normal model for the same sentences. The normal model
will be of the same size or smaller than the one we started with (as can easily
be seen by looking at what we were doing), so we can therefore conclude that
the proof of the Gödel–Henkin theorem shows that every consistent set of
sentences in a countable language with an equality predicate letter which
includes the equality axioms has a countable normal model. As an example,
consider the structure $\mathscr{R} = \langle R, <, +, \cdot \rangle$ where R is the set of real numbers,
$<$ is 'less than' as before, and $+$ and $\cdot$ are the ordinary plus and times functions defined on the real numbers. Now consider an appropriate language
(which is countable) for $\mathscr{R}$. (For example, it might have the symbols P, f,
and g interpreted as $<$, $+$, and $\cdot$.) Consider the set of sentences of this
language that are true in $\mathscr{R}$. Call this set of sentences Σ^R. Since Σ^R is a set of
formulae from a countable language, Σ^R must have a countable normal model,
say $\langle A, \tilde{<}, \tilde{+}, \tilde{\cdot} \rangle$, where A is a countable set. It seems peculiar in the
least to say that the same set of sentences is true of both $\mathscr{R}$ and $\mathscr{A}$, especially
if we take, as we can, A to be a set of reals (that is, A is a countable set and
its members are real numbers). It means, for example, that whenever we
describe a real number as the unique number having some property defined
by a formula in the formal language, this number is already in A. The remaining
reals are thus in a sense redundant. Moreover, this fact might at first sight
seem to contradict various characterizations of the real number system
(although of course it does not). Another, perhaps clearer, example of this
oddity comes from set theory. Consider the structure $\mathscr{S} = \langle$all sets, is a
member of$\rangle$. The set of sentences that are true in this structure must also

have a denumerable model. One such sentence, in fact derivable from the Power Set Axiom, (*see* Chapter 6) says that there is a set with more than a denumerable number of members in it and we shall see later in the book that the axioms imply that there are very many sets bigger than the set of natural numbers. But if the axioms of set theory have a denumerable model, then these are sentences which will be true in this denumerable model. And this too is somewhat surprising. How can a sentence saying there is a set with uncountably many members in it be true when there are only countably many things altogether? What we have to do in order to resolve this paradox is to look carefully at what the sentences do in fact say.

In our denumerable model, every infinite set is *in fact* denumerable, so there is a one-to-one correspondence between it and the natural numbers. But this correspondence, when expressed as a set, may not belong to the denumerable model, which shows how it may contain a denumerable set which nevertheless satisfies *in the denumerable model* the predicate 'is an uncountable set'.

Before I tie all this together, I want to present a theorem, due to Löwenheim, Skolem, and Tarski, which shows that not only can we get smaller models, but also larger ones:

THEOREM: If a set Σ of sentences in a countable language has an infinite normal model, then Σ has normal models of all infinite cardinalities. That is, for any infinite set S, there is a normal model $\langle A, \ldots \rangle$ of the set of formulae Σ such that A has the same cardinality as S, if indeed Σ has any infinite normal model.

So from this theorem we know that any set of sentences which are true in some infinite structure cannot even tell us the size of the underlying set. If a set of sentences has a model of infinite cardinality then we are unable to deduce this cardinality or anything about it from the set of sentences (except that it is infinite).

So far all I have done is to use model theory to point out some things about what sets of sentences *cannot* say about their models. We know that if a set of sentences has a model then it has a countable normal model, and we cannot in general say that these are the same. We know that if a set of sentences has an infinite model then it has a model of every infinite cardinal. We also know in particular that a set of sentences, Σ say, has models where the underlying sets are sets of numbers and models where they are not. But what positive things can we say? What does it mean when we can find a set of sentences, say Σ, which is such that, for some particular model $\mathcal{A}$, $\mathcal{A} \models \psi$ if and only if $\Sigma \vdash \psi$?

Well, there is an infinite class of questions of the form: is a sentence ψ true in $\langle N, < \rangle$? There is a question like this for every ψ. One way of answering this question is to axiomatize the set of truths (which the set Σ we have given does) and then to attempt to discover a proof of the ψ in question. Since for every ψ, either ψ or $\neg\psi$ is true in $\langle N, < \rangle$, one or the other is provable from Σ. If we are sufficiently systematic in our search for such a proof, we can guarantee that we shall find one or other eventually, which will establish whether ψ is true or not in $\langle N, < \rangle$. Thus by proving our lemma, we establish a systematic method for answering these questions. So we have this much: we know exactly which sentences are true in $\langle N, < \rangle$. We can, with a little more difficulty, do the same for the structure $\langle N, <, + \rangle$ that we have done for $\langle N, < \rangle$. That is, we can write down a fairly obvious set of sentences true in this structure, and then using model-theoretic methods show that from these sentences everything true in this structure can be proved. In this way we get a practical method for finding whether a sentence is true in the structure or not. The same procedure can be applied to the real and complex number fields, although not to the structure $\langle N, <, +, \cdot \rangle$, as Chapter 5 will show.

4
Turing Machines and Recursive Functions

As can be seen from other chapters in this book, mathematical logic is frequently concerned with infinite classes of statements, and seeks to deal with them in systematic, uniform, or mechanical ways. For example, in Chapter 3 a systematic method was found for determining the truth of a given sentence in the structure $\langle N, < \rangle$, and in Chapter 2 a systematic method was found for listing all the true statements of predicate calculus. In both cases the method was clearly reducible to a mechanical computation, in some sense, so these may be regarded as positive results in some 'theory of computability'.

In other cases, such as deciding whether a given statement of predicate calculus is true (as distinct from finding it in the list of theorems if it *is* true) we lack systematic mechanical methods, so here are possible candidates for *negative* results in the theory of computability, i.e. there may be proofs that no mechanical solution is possible. The trouble with the search for negative results is that it necessitates a definition of computability—as long as we are exhibiting actual computation methods in our positive results no definition is necessary, but to show that *no* mechanical method solves a given problem presupposes a definition that embraces all types of computation.

To embrace all forms of computation naturally leads to a certain amount of intricacy; however, the main lines of this chapter are basically straightforward and almost inevitable—the definition of computability leads to a problem unsolvable by computation, then the universal expressive power of the predicate calculus leads to the translation of this problem into logic, and thence to the general unsolvability of the problem of logical validity.

Turing machine computation

Turing and Post independently in 1936 arrived at a precise analysis of the concept of computation. Since the notion of computation is an intuitive one, any precise formulation of this notion will have to rest on evidence rather than mathematical proof. However, we can say that no evidence has so far been found to shake our confidence that Turing machines, as they are called, can perform all possible computations. Turing's analysis in addition exposes the reasons why no computation procedure, in the sense of an unambiguous set of instructions which a human being could follow without using any imagination, is likely to be one which a Turing machine could not carry out.

A Turing machine has a tape which is endless in both directions. The tape is divided into squares, thus

and each square can have a symbol written on it. The tape is inspected or *scanned* one square at a time by a *reading head*. The actual technology involved in a machine with a reading head does not concern us. A given machine has a specific *alphabet* consisting of a finite number of symbols $S_0, S_1, \ldots, S_n$ (and we take S_0 to be the blank square $\square$). In addition the machine possesses a number of *internal states* $q_0, q_1, \ldots, q_m$. The machine's act at a given time is uniquely determined by its internal state and the symbol in the scanned square, and this act will be to either

(i) change the scanned symbol,

(ii) move one square to the right,

or

(iii) move one square to the left.

A machine is completely specified by a finite table of quadruples of the following kinds:

	State	Scanned symbol	Act	Next state	
(i)	q_i	S_j	S_k	q_l	(replace symbol)
(ii)	q_i	S_j	R	q_l	(move right)
or					
(iii)	q_i	S_j	L	q_l	(move left).

We shall show how such a machine operates shortly.

The unique determination of machine operations will be reflected by the fact that no two quadruples commence with the same ⟨state, symbol⟩ pair. If the machine ever reaches a ⟨state, symbol⟩ combination for which no quadruple exists, then it halts. If the machine is in state q_k scanning a square thus

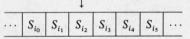

we shall denote this situation by

$$\ldots S_{i_0} \; S_{i_1} \; q_k \; S_{i_2} \; S_{i_3} \; S_{i_4} \; S_{i_5} \ldots .$$

For example, suppose the machine has instructions $q_1 \ S_1 \ L \ q_2$ and $q_2 \ S_2 \ L \ q_2$ and suppose the non-blank part of the tape has on it

$$S_1 \ S_2 \ S_2 \ S_1 \ S_2 \ldots S_1,$$

and the machine is in state q_1 scanning the second S_1, that is, we have

$$S_1 \ S_2 \ S_2 \ q_1 \ S_1 \ S_2 \ldots S_1.$$

Then the instruction $q_1 \ S_1 \ L \ q_2$ can be implemented and we get the situation

$$S_1 \ S_2 \ q_2 \ S_2 \ S_1 \ S_2 \ldots S_1.$$

Now the second instruction can be implemented to get

$$S_1 \ q_2 \ S_2 \ S_2 \ S_1 \ S_2 \ldots S_1.$$

This can be repeated, thus

$$q_2 \ S_1 \ S_2 \ S_2 \ S_1 \ S_2 \ldots S_1,$$

and, as no instruction starts with $q_2 \ S_1$, the machine now halts.

When constructing a machine table it is helpful mentally to put oneself in the machine's position of being able to see only one square at a time. Internal states correspond to 'mental notes' of situations previously scanned and hence enable 'memory' of one part of the tape to be carried to another part.

In practice tables rapidly become lengthy even for simple computations, so we develop a stock of basic machines which perform tasks involved in a variety of computations. Then in constructing a complex machine we replace the lines in its table by tasks for which machines have already been constructed, since it is a routine matter to arrange the suffixes on the q's so as to link up basic machines. For example, if the first machine uses $q_1, \ldots, q_{20}$ and the second $q_1, \ldots, q_{12}$ then we renumber the q's of the second machine as, say, $q_{21}, \ldots, q_{32}$.

Examples of basic tasks are

1. Search to right for S_j.

State	Scanned symbol	Act	Next state
q_0	S_0	R	q_0
q_0	S_1	R	q_0
	$\vdots$		
q_0	S_{j-1}	R	q_0
q_0	S_j	S_j	q_1
q_0	S_{j+1}	R	q_0
	$\vdots$		
q_0	S_n	R	q_0

This machine halts on S_j. If we want the machine to take further action, we can use q_1 as the first state in a further series of quadruples. For example, a repeat of the above with q_1 in the place of q_0 and with q_2 in the place of q_1 gives a search for the *second* S_j to the right of the starting point. A 'search left' machine is of course similar.

2. Put a marker on S.

$$q_0 \quad S \quad S' \quad q_0$$

Similarly, $q_0\ S'\ S\ q_0$ 'removes the marker' from S. Although we have to treat S' as a single symbol, all the advantages of a movable marker can be obtained. To put a marker on any square simply replace the symbol S in that square by S'.

3. Move right, erasing all markers.

$$\left.\begin{array}{cccc} q_0 & S & R & q_0 \\ q_0 & S' & S & q_0 \end{array}\right\} \text{ for all symbols } S,\ S' \text{ in the machine's alphabet.}$$

If we want this to stop at a given symbol, say □, we omit those quadruples whose second entry is □.

(Markers are an indispensable item of Turing machine technique in one form or another. The basic reason is that any machine has its internal memory, that is, the number of internal states, bounded in advance, whereas general computation requires unbounded amounts of memory. For example, to compare the lengths of two blocks of symbols on its tape a machine cannot rely on internal states alone. If a machine has n internal states there will be two blocks among those of length at most $n + 1$ for which the machine arrives at the same internal state after traversing them both. Therefore it can only decide between them on the basis of marks made on its tape. The most obvious way to do this is to zig-zag back and forth between the blocks, marking one more square each time a block is visited, until one block is exhausted. (See also the example below.))

4. Move each symbol on the right of the scanned square one square to the right.

This machine requires two special states for each symbol S_i. Accordingly we use letters as subscripts for these states, rather than numerals, to show their role more clearly. Firstly, a 'remember S_i' state which we call q_{RS_i} which the machine enters when it sees the symbol S_i and secondly a 'deliver S_i' state q_{DS_i} which the machine assumes as it leaves the square formerly bearing S_i.

Then for each pair of machine symbols S_i, S_j we list the quadruples

$$q_0 \quad S_i \quad S_i \quad q_{RS_i}$$
$$q_{RS_i} \quad S_i \quad R \quad q_{DS_i}$$
$$q_{DS_i} \quad S_j \quad S_i \quad q_{RS_j}$$

If we want the machine to stop when it reaches some end marker S_k then we change the q_{DS_k} quadruple to make a transition into some halting state.

Now as an example of the way a complex task may be synthesized from basic tasks, here is a machine which doubles a given block of 1's. The diagram represents the initial machine situation, followed by the situations which result when the quadruples and basic tasks are applied. The machine copies one square at a time into the blank portion of its tape, putting a marker on the symbol to be copied before it does so. Then when the machine finds no unmarked square left in the given block, it knows it has copied the entire block, so it erases the markers, and halts.

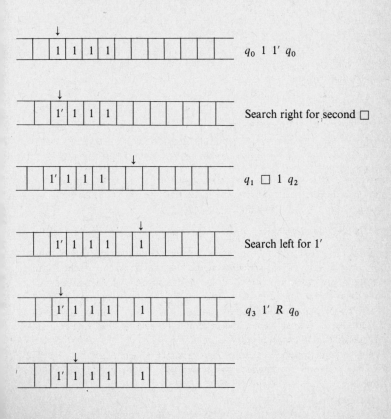

The cycle then repeats to give

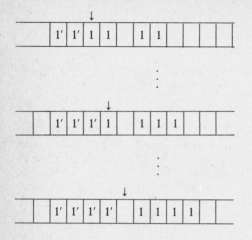

At this point the machine is in state q_0 scanning $\square$, a situation for which no quadruple has yet been given, so we now have the instruction 'Move left, erasing markers' and after applying it get:

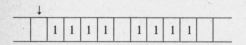

Expanding the basic tasks into their tables of quadruples, and arranging state subscripts so as to link up the successive basic tasks, we can obtain the standard table for this machine.

Partial recursive functions

Let us take a block of $n + 1$ ones as the standard way of presenting a number n to a machine (we need one mark for 0 just so the machine knows it is being presented with something). Then a partial function ϕ is associated with each Turing machine M, where $\phi(n)$ is the number of ones left on M's tape after M has completed a computation beginning with presentation of the number n. We say 'partial' because not every n may lead to a completed computation—M may not halt.

A k-tuple of numbers (a set of k numbers in order) can be presented as k blocks separated by single blank squares, and a k-place partial function associated with M similarly.

Definition. A partial function ϕ is called *partial recursive* if ϕ can be associated with some Turing machine M in the above manner; ϕ is called *recursive* if it is defined for all arguments (that is, is total).

Thus if we accept that all computations can be done by Turing machines then the computable partial functions are precisely the partial recursive functions. (We are forced to consider partial functions, rather than total functions only, because there is no computable method for filtering out the partial functions which are not total. This will be apparent later in our discussion of the halting problem.)

Our machine example above shows that the function $\phi(n) = 2n$ is partial recursive, and it is easy to construct machines which show that $\phi(m, n) = m \times n$ and other common functions are partial recursive.

Connected with the notion of computability is the notion of *solvability of problems by an algorithm*. The kind of problem I refer to is that where a uniform method is sought to answer an infinite class of questions Q. For instance, Q might be: is c the greatest common divisor of a and b (for natural numbers a, b, c)? As we know, the Euclidean algorithm is available to solve this problem, and it is easy to construct a Turing machine M which will take any 3-tuple $\langle a, b, c \rangle$ and eventually signal YES (say by stopping when it is scanning a square with a 1 on it if Q has answer YES), and signal NO (say by stopping on $\square$) otherwise.

In general, if the questions Q possess some natural representation in some finite alphabet, so that they can be put on the tape of a Turing machine, then we may set up the following definition.

Definition. The class of questions Q is solvable (or decidable) if there is a Turing machine M which, when applied to any Q, eventually stops on 1 if the answer to Q is YES, and on $\square$ if the answer to Q is NO.

Standard description of Turing machines

In this section we lay the foundation for an argument which is often called 'diagonal' or 'self-referential'. Cantor's proof of non-denumerability of the real numbers and Gödel's sentence which asserts its own unprovability (see next chapter) are both examples of this type of argument. The diagonal argument is often regarded with suspicion; however in reality it is perfectly concrete and non-paradoxical. In the case of Turing machines it rests upon the fact that any Turing machine can be described by a finite sequence of symbols, and finite sequences of symbols are what Turing machines themselves act upon.

Since each Turing machine requires only finitely many symbols we can assume without loss of generality that the symbols are drawn from the list $\square, 1, 1', 1'', 1''', \ldots$. State symbols can also be chosen from the list $q, q', q'', q''', \ldots$. Then if we regard the foregoing rather as strings of symbols built using $\square$, 1, q, and $'$ it is clear that any quadruple can be written using only the symbols $\square, 1, ', q, R, L$ and hence we can represent any Turing machine as a word in this six-letter alphabet by stringing quadruples together. So, for example, the machine

$$q_0 \quad 1 \quad R \quad q_1$$
$$q_1 \quad 1'' \quad 1' \quad q_2$$

would be represented by the word

$$q1Rq'q'1''1'q''.$$

Such a representation is unambiguous and can be used as input for a Turing machine. However, since we agreed to restrict machines to operate on the symbols $\square, 1, 1', 1'', \ldots$ we must first code our six letters into the standard alphabet. We do this as follows:

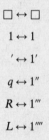

$$\square \leftrightarrow \square$$
$$1 \leftrightarrow 1$$
$$' \leftrightarrow 1'$$
$$q \leftrightarrow 1''$$
$$R \leftrightarrow 1'''$$
$$L \leftrightarrow 1''''$$

We call the word in this alphabet which is associated with the machine M the *standard description* of M and denote it by $\ulcorner M \urcorner$.

An unsolvable problem

Consider the infinite class of questions (one form of the so-called 'halting problem').

$$Q_M: \text{ Does } M \text{ applied to } \ulcorner M \urcorner \text{ eventually stop on } \square?$$

It is reasonable to assume the question Q_M to be represented by the word $\ulcorner M \urcorner$, since this word contains all the necessary information. A machine S which solves this problem would then take the word $\ulcorner M \urcorner$ and eventually stop on 1 if the answer to Q_M was YES, or on $\square$ if the answer to Q_M was NO.

But how does S act on $\ulcorner S \urcorner$? If S stops on 1 this indicates that the answer to Q_S is YES, that is, that S stops on $\square$ after being applied to $\ulcorner S \urcorner$. And if S stops on $\square$ this indicates that the answer to Q_S is NO, in other words that S does not stop on $\square$. This contradiction shows that the machine S does not exist, and hence there is no general method for solving the problem represented by the class of questions Q_M and we say that Q_M is algorithmically unsolvable.

It may be claimed that the set-up of the problem unfairly exploits our convention for the signalling of YES and NO by machines. However any other machine T which solves the problem under another convention can obviously be converted into a machine which solves the problem under the agreed convention—simply by linking T to a machine which translates T's YES into a 1 and T's NO into a $\square$. Such a combination machine cannot exist by the argument above; hence neither can T.

The same argument applies to the other end of the problem—the way in which Q_M is expressed. For any reasonable expression of Q_M there is a Turing machine which converts $\ulcorner M \urcorner$ into this expression, so linking this to the hypothetical solution in terms of the new representation again gives a solution which we have shown to be impossible.

In particular, the problem could be given a purely numerical representation—interpret $\ulcorner M \urcorner$ as a numeral in base 6 and try to calculate

$$\psi(M) = \begin{cases} 1 \text{ if answer to } Q_M \text{ is YES,} \\ 0 \text{ if answer to } Q_M \text{ is NO.} \end{cases}$$

Since no machine can do this, ψ is a function which is not partial recursive.

A further source of uneasiness with this problem may be expressed as follows: of what mathematical significance are such undecidability results if they all rest on pathological 'self-reference' constructions? The answer to this is that the same pathology actually occurs in such superficially healthy structures as predicate calculus and group theory. It is possible to show (as we shall for predicate calculus) that a solution of the problem of deducibility in either of these theories would yield a solution of the problem of the Q_M's.

Universal machine

It may be thought that the problem of the Q_M's is unsolvable because of the inability of any single machine to encompass the workings of all machines. In fact, a universal machine U does exist, a machine which will take the standard description $\ulcorner M \urcorner$ of any machine and a coding $\ulcorner P \urcorner$ of any tape pattern P, and proceed to simulate the action of M applied to P. So we are forced instead to the conclusion that an unsolvable problem exists concerning U.

U, as we define it below, will simulate the action of M applied to $\ulcorner M \urcorner$ if given the word $*\ulcorner M \urcorner**\ulcorner M \urcorner*$. (We use $*$'s as markers so that we can detect the ends of the relevant portion of tape and as separation between the machine description and the tape data.) Hence there can be no algorithm for deciding whether U eventually stops on $\square$ after being applied to a word of this form, since $\square$ is coded by $\square$ and therefore M stops on $\square$ if and only if the simulation does also. Consequently there can be no algorithm to decide the broader class of questions:

Q_W: Does U applied to a word W eventually stop on $\square$?

This 'condensation' of the problem of the Q_M's to a problem about a single machine is crucial to the undecidability proof for the predicate calculus, for we shall ultimately code the machine U as a formula ϕ of predicate calculus, and express the effect of U on given inputs as logical consequences of ϕ.

Now the existence of a universal machine is apparent as soon as one realizes that there is an algorithm for reproducing the effects of a given machine M on a tape pattern P. Markers are used to tag the current internal state among the quadruples of M and the square in P currently being scanned. It is a simple matter to move back and forth between them, carrying out the change in P indicated by the quadruple which is being applied, finding the quadruple which commences with the next state, shifting the current state marker to it, etc.

A technical complication arises from the fact that U has only finitely many internal states and hence it can only read and 'remember' finitely many different symbols. This means that U must operate with blocks of symbols of different lengths to represent different states and symbols in the machine M being simulated, and laboriously compare blocks square by square to see if they actually represent the same symbol. Also, to replace one symbol by another U must actually replace one block by another, generally of a different length, so portions of U's tape pattern must be shifted one square to the right or left repeatedly until the new block being created finds a space in which it fits.

All these operations can be accomplished by synthesis from basic tasks such as 1–4, and it is preferable to give a diagram of the underlying structure of U rather than attempt to list hundreds of quadruples.

As stated above, U is applied to a tape of the form $*\ulcorner M \urcorner**\ulcorner P \urcorner*$, and markers are kept on the current state block in M and the scanned symbol block in P. (The $*$'s should of course be some symbol in the standard alphabet, for instance $1''''$.) The current state block will be followed by the 'current symbol block', 'current act block' and 'next state block'. U can then be described by the following flow chart:

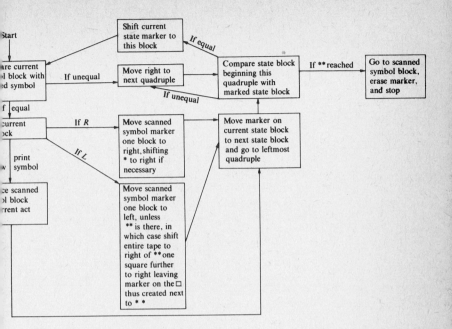

Turing machines in terms of word transformations

Since we are concerned with machines whose tape contains only finitely many marks at a given time all the essential information in a given machine situation can be encoded in a word which includes (i) the sequence of symbols on the portion of the tape which includes the scanned square and all marks and (ii) the position of scanned square and current state. One way of doing this is illustrated by the following example. Represent

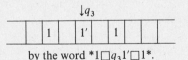

by the word $*1\square q_3 1' \square 1*$.

(The state symbol is placed to the left of the symbol representing the scanned square, and *'s are used as end markers.)

If we call such a word W a situation word, then the word W' for the immediately consequent situation will be determined by a transformation involving the combination $q_i S_j$ occurring in W. The following transformations of words accomplish the change from W to W' in all cases.

Type of quadruple				Transformation	
q_i	S_j	S_k	q_l	$q_i S_j \rightarrowtail q_l S_k$	
q_i	S_j	R	q_l	$q_i S_j S_k \rightarrowtail S_i q_l S_k$	(for each S_k)
				$q_i S_j^* \rightarrowtail S_j q_l \square^*$	
q_i	S_j	L	q_l	$S_k q_i S_j \rightarrowtail q_l S_k S_j$	(for each S_k)
				$^* q_i S_j \rightarrowtail {}^* q_l \square S_i$	

The $*$ in effect makes new blank squares available whenever they are needed for further movement of the q symbol to the right or the left.

A given machine M is thus representable by a finite set of word transformations corresponding to its quadruples. Given any situation word, at most one of these transformations applies to it, and the sequence of words obtained correctly reflects M's sequence of situations. In particular, a situation where M halts on $\square$ corresponds to a word containing a combination $q_h \square$ not occurring on the left hand side of any transformation for M.

For each such combination let us add to the transformations for M

$$q_h \square \rightarrowtail \diamond \quad (\diamond \text{ a new symbol})$$

and

$$\left. \begin{array}{c} \diamond S \rightarrowtail \diamond \\ S \diamond \rightarrowtail \diamond \end{array} \right\} \text{for any other symbol } S.$$

Thus the occurrence of a word on which the machine M halts creates a symbol $\diamond$ which swallows all other symbols, leaving only the word $\diamond$ itself. Unlike the transformations for M the latter transformations are not completely deterministic (though this could be arranged), because more than one may be applicable to a given word; however they can come into operation only if the initial situation word leads to a halting on the $\square$ situation.

Let us call this extended set of transformations the M-calculus and write $W_1 \rightarrowtail W_2$ in general if there is a sequence of transformations in the M-calculus which takes W_1 to W_2. Then if W is a situation word we have $W \rightarrowtail \diamond$ if and only if the situation described by W eventually leads to M halting on $\square$, so when M is a machine (such as U) for which the latter problem is unsolvable we obtain the following unsolvable problem:

Decide for any given word W, whether $W \rightarrowtail \diamond$ via the transformations of the U-calculus.

Representation of word transformations in predicate calculus

In this section the halting problem is finally translated into a problem of deducibility in the predicate calculus, using the above problem of transformation in the U-calculus. We use a language which has each symbol $\square$, 1, $\diamond$, $*$

etc. of the U-calculus as a constant, a function f which links words together, and a single binary predicate Tr for transformation.

As we know, constants and function letters can in principle be eliminated from the predicate calculus, as can the rule which allows substitution of constant terms for variables. I use them because they make the parallel between word transformations and deductions more transparent.

1. Function for linking letters.

Consider a function $f(x, y)$, which we write (xy) for simplicity, governed by the axiom

(1) $$(x(yz)) = ((xy)z).$$

Any word can then be regarded as a constant term, for instance $*\square q_3 1*$ is $(*(\square(q_3(1(*)))))$, and axiom (1) allows us to rearrange the brackets any way we please, so brackets are actually immaterial and we omit them from now on.

2. Axioms for transformation.

We can consider $t_1 \rightarrowtail t_2$ for any terms t_1, t_2 as a formula of predicate calculus. (The reader may represent it more conventionally as $\mathrm{Tr}(t_1, t_2)$ if he prefers.) When t_1, t_2 are constant terms W_1, W_2 the formula $W_1 \rightarrowtail W_2$ asserts that W_1 is transformable into W_2 by the U-calculus.

We want to write down enough axioms to guarantee the deducibility of the formula $W_1 \rightarrowtail W_2$ if and only if $W_1 \rightarrowtail W_2$ actually holds.

First, write

(2) $$xTy \rightarrowtail xT'y$$

for each transformation $T \rightarrowtail T'$ in the U-calculus. Then if W' is an immediate consequent of W in the U-calculus we can deduce $W \rightarrowtail W'$ as follows: W must be of the form

$$W = XTY$$

for some words X, Y and some T which occurs as a left-hand side of a transformation in the U-calculus. But then

$$W' = XT'Y$$

and we can deduce $XTY \rightarrowtail XT'Y$ from (2) by substitution of the constant terms X, Y for the variables x, y. It is also clear that all relations $W \rightarrowtail W'$ deducible from (2) correctly represent immediate consequences in the U-calculus.

Now we only have to add

(3) $$(x \rightarrowtail y \mathrel{\&} y \rightarrowtail z) \rightarrow (x \rightarrowtail z)$$

to be able to prove $W_1 \rightarrowtail W_2$ if W_2 is any word into which W_1 can be transformed, and again it is clear that any deductions from (3) correctly represent actual word transformations.

Thus we have

$$((1) \ \& \ (2) \ \& \ (3) \vdash W_1 \rightarrowtail W_2) \text{ if and only if } W_1 \rightarrowtail W_2.$$

3. Elimination of axioms; undecidability.

Since there are only finitely many transformations $T \rightarrowtail T'$ in the U-calculus, there is a single formula ϕ which is the conjunction of (1), (2), and (3). Then

$$W_1 \rightarrowtail W_2 \text{ if and only if } \phi \vdash W_1 \rightarrowtail W_2 \text{ if and only if } \vdash \phi \rightarrow (W_1 \rightarrowtail W_2).$$

In particular, for any word W,

$$W \rightarrowtail \diamondsuit \text{ if and only if } \vdash \phi \rightarrow (W \rightarrowtail \diamondsuit).$$

Given any W, we can effectively construct the formula $\phi \rightarrow (W \rightarrowtail \diamondsuit)$, so if we could decide whether this formula was deducible we could decide whether $W \rightarrowtail \diamondsuit$.

As established in the last section, the latter problem is unsolvable, hence so is the problem of deducibility in predicate calculus. That is, *there is no algorithm which will enable us to decide, given any formula ψ of predicate calculus, whether or not ψ is deducible from the axioms of predicate calculus.*

By the completeness theorem, the deducible formulas in predicate calculus coincide with those that are logically valid, so this result also shows that no algorithm exists for deciding whether a given formula is logically valid.

5

Gödel's Incompleteness Theorems

EARLIER this century the mathematician Hilbert posed the problem of finding a formal system which gave all the true mathematical statements, and only those. This was called 'Hilbert's programme'. However, the simple case of formal arithmetic disrupted this programme. By formal arithmetic I mean a formal system which will deal with the arithmetic of the natural numbers $0, 1, 2, \ldots$ and the usual basic functions such as addition and multiplication. In 1931, Gödel proved that if a formal system, let us call it F, included arithmetic, then (i) there is a statement of F (or indeed of arithmetic) which is true, but not provable and (ii) we need a system stronger than F if the consistency of F is to be proved.

In this chapter I am going to describe a formal system of arithmetic which will be strong enough to treat all the usual things in arithmetic that one might want to deal with, including even such exotic things as Fermat's Last Theorem. Also in this chapter I shall exhibit a formula which says 'I am not provable'. This formula will be true, so clearly it will not be provable.

In fact it is very easy to describe this formula. I shall go into the detail later. For the moment, consider a formula which I shall abbreviate as $\exists x \, \mathrm{Pf}^+(x, b, c)$. This is to be true only when x is the Gödel number (and I shall say what this means later on) of a proof of the formula that one gets when one substitutes the numeral $\bar{c}$ (I shall say what this means later on, too) for the free variable y, in the formula whose number is b. When this formula is interpreted in the ordinary integers, it says that there is an x such that x is the Gödel number of a proof of a certain formula when $\bar{c}$ is substituted for its only free variable. The number of the formula is b before this substitution is performed.

Now consider the formula $\neg \exists x \, \mathrm{Pf}^+(x, y, y)$. This formula itself will be given a Gödel number g, say. We shall investigate what happens when you put in the numeral $\bar{g}$ of the number g in the formula for the y. If you examine this formula (and I shall reiterate this later), what it says is that there is no proof of this very same formula and that is indeed true; there is no proof of this formula because if there were a proof, we should have a contradiction. Indeed one can show that there is neither a proof of it nor a proof of its negation. What we shall have then, when we have filled in the details, will be a formula which is true but not provable. And there are further ramifications. One can actually show that in arithmetic one cannot prove the consistency of arithmetic and the argument for this depends upon a formalization of this former argument.

This is what I am going to lead up to. Now I want to go right back to the beginning and consider basic arithmetic. What are the crucial features of arithmetic? We want to be able to talk about zero, and, given any particular number, we want to be able to add 1 to it. There is one other important thing that we want to be able to talk about: mathematical induction. Mathematical induction, intuitively speaking, says that if zero has a certain property and if, whenever a particular individual number n has that property, $n + 1$ also has that property, then every number has that property. This is something which is probably familiar from high-school mathematics. In writing down formulae it is pretty obvious that some can be produced that deal with numbers and induction and as I remarked earlier, if you have any formal system that involves any reasonable formal system for arithmetic, then Gödel's Incompleteness Theorem applies. So we shall now describe quite a simple formal system.

It is going to be in a language with a constant symbol 0 interpreted as zero. It has a one-place function letter s which I shall read as 'successor'. It is interpreted as adding one. We shall have also $+$, $\times$, and $=$, interpreted as plus, times, and equals. Apart from this we shall have all the usual machinery of predicate calculus. And we could have lots of other things too. We could even have variables ranging over functions and so on; it will not make any difference at all. The proof will still go through. Indeed, when I get to the question of axioms, we shall see that we can throw in a few extra of these too. There will be certain restrictions here: if you throw in an infinite number of axioms, then you might not be able to prove Gödel's theorem. For example, if you add as axioms *all* the statements of arithmetic in our formal language which happen to be true, then everything which is true is provable and everything which is provable is true, so that is going too far. But if we just add any finite number of axioms or schemes for axioms, then we shall be able to carry out Gödel's construction.

So what do we take as our axioms for arithmetic? In fact the following will suffice. First we have axioms for equality. Those have been touched upon before in Chapter III. Essentially all we need are

(i) $x = x$,

(ii) $x = y \rightarrow (x = z \rightarrow y = z)$,

(iii) $x = y \rightarrow (A(x, x) \rightarrow A(x, y))$, for any formula A of the language with two free variables. This says if x and y are the same, then any property of one is a property of the other and this is so because we are thinking of x and y as being identical objects. So these are just the ordinary axioms for equality.

What are the special axioms we need? Peano wrote down a set of axioms in informal mathematics which exactly characterize the natural numbers. It turns out that just writing down an analogue of these things is quite sufficient for our purposes. The first axiom says that 0 is not the successor of anything.

$$\text{(a)} \quad \neg 0 = sx.$$

The next one says that given any number, its predecessor is unique.

$$\text{(b)} \quad sx = sy \rightarrow x = y.$$

And now we need axioms that tell us how plus and times behave:

$$\text{(c)} \quad x + 0 = x,$$
$$\text{(d)} \quad x + sy = s(x + y),$$
$$\text{(e)} \quad x \times 0 = 0,$$
$$\text{(f)} \quad x \times sy = x \times y + x.$$

I am going to write them in this way. Formally one ought to write, for example, $+(x, y)$ but for convenience I write $x + y$. It is merely a technical trick that is involved here.

We also want to include an induction axiom, but there is no need to include any other axioms governing other functions because all the ones that I think anybody could reasonably think of can be obtained from these axioms straightforwardly. But the induction axiom is rather special. It is not really an axiom, of course. It is a scheme. Given any formula with a free variable x, say $P(x)$, we take as an instance of the induction axiom scheme:

$$\text{(IS)} \qquad (P(0) \,\&\, \forall x(P(x) \rightarrow P(sx))) \rightarrow \forall x P(x).$$

The first question we ask is: Is this system powerful enough to give us all the things we want? The answer is: 'Yes, it is'. I shall not ask for challenges from the reader. Just as an example we can express x divides y in this system simply by the formula $\exists z(x \times z = y)$. And it is very easy to write down a formula which says that x is a prime number by just talking about the numbers which divide x.

We also have representations of all the numbers $0, 1, 2, \ldots$. These are represented in the formal system by the *numerals* $0, s0, ss0, \ldots$ respectively. We write $\bar{n}$ for a string of n s's followed by a 0.

Now I am glossing over things a bit by writing this. If it is not clear we can be thoroughly pedantic by writing, say, $s(s(0))$ instead of $ss0$. And this will be the representation of 2. If we wanted to write $2 + 2 = 4$ in this way, we should get

$$(*) \qquad s(s(0)) + s(s(0)) = s(s(s(s(0)))).$$

Now what I was talking about earlier was representing a formula by numbers. Gödel found a way of doing this by what is now known as Gödel numbering. We have a look at our language and what we do is to assign numbers to all the basic symbols in our language. What are the basic symbols? We have 0 and we give this the number 1 which is the Gödel number zero. (We want Gödel numbers to be non-zero for technical reasons which will emerge later on.) s gets the number 2, + the number 3, and so on:

0	1
s	2
+	3
×	4
=	5
(	6
)	7
,	8
x	9
|	10
⌐	11
&	12
∃	13

I want to have lots of variables $x_1, x_2, \ldots$. A convenient device is to put little bars as subscripts to x to get: $x_|, x_{||}, x_{|||}, \ldots$ and we shall use these instead. The list gives all the symbols that we have in our language. If we had others, we could give them numbers, too. It is convenient to use only a finite number of symbols, which is why I put the bars on the x for the variables, but it is not essential.

If you take the rather ghastly expression (*) above for $2 + 2 = 4$ in the form $+ (s(s(0)), s(s(0))) = s(s(s(s(0))))$ and write down the corresponding numbers, it comes out to be

3 6 2 6 2 6 1 7 7 8 2 6 2 6 1 7 7 7 5 2 6 2 6 2 6 2 6 1 7 7 7 7

Now that one is all right. It is quite clear how to recover the $2 + 2 = 4$ from this mass of numbers. But what happens if we have something like 139? Should we interpret this as being a 13, which is the number of an ∃, followed by a 9, which is the number of x, and thus together stands for ∃x, or should we interpret it as a 1 followed by a 3, followed by a 9 which is the sequence of numbers for $0 + x$? There is in this case no way to decide how to interpret the number. We need some device that enables us to distinguish between two or more possible interpretations. Now suppose we take a simple number, for example 9 720 000 000. We can write this as $2^9.3^5.5^7$ and we can write it in only one way when the bases are the primes in order of magnitude. Now if we regard $2^9.3^5.5^7$ as coding the sequence of numbers 9, 5, 7, we are safe because we can code a sequence in exactly one way by using this device. Thus we can uniquely get the number code for $x = 0$. So if we use the prime numbers in this way, we can code into a single number a whole string of numbers, and given a number, we can always factorize it and recover the expression of our language in a unique way. The numbers will get very large indeed, but this is of no theoretical significance. For example, the number we get for the formal expression (∗) for $2 + 2 = 4$ is: $2^3.3^6.5^2.7^6.11^2.13^6.17^1.19^7.$ $23^7.29^8.31^2.37^6.41^2.43^6.47^1.53^7.59^7.61^7.67^5.71^2.73^6.79^2.83^6.89^2.97^6.101^2.103^6.$ $107^1.109^7.113^7.127^7.131^7$. This, when explicitly calculated, is rather large but it does not matter because in principle we can always calculate these numbers and mathematics is only concerned with principles and not with the feasibility of working out some numerical expression. As another example, suppose we want to code $\neg x = 0$. Now $\neg$ gets 11, x gets 9, = gets 5, and 0 gets 1, so the Gödel number for the formula is $2^{11}.3^9.5^5.7^1$. I shall write $\ulcorner \phi \urcorner$ for the Gödel number of ϕ, so $\ulcorner \neg x = 0 \urcorner = 2^{11}.3^9.5^5.7^1 = 881\ 798\ 400\ 000$. Of course there are some numbers which do not code formulae, for example the number 3, since $3 = 3^1 = 2^0.3^1$ which codes 0, 1 which is not a sequence of Gödel numbers corresponding to a formula of the language. But that is all right, because we can always tell whether we code anything or not because all the exponents have to lie between 1 and 13. These are the only ones which we have been using.

I said that we could represent the numbers by numerals. That is, we can represent zero by 0, one by $s(0)$, two by $s(s(0))$, etc. Now I want to know whether we can have a machine which tells us whether n is the Gödel number of a numeral or not. Or alternatively we could ask: What are the possible Gödel numbers of the numerals? If something is a numeral, one possibility is that it could be zero and the other is that it starts off with an s, which has Gödel number 2, and then has a pair of brackets around something which is already a numeral. Now we can write down a machine for this.

The crucial point of course is that all these things are computable. You really can have a machine that will decide in a finite time whether n has this property or not. Now numerals, which are expressions in the language, are defined by saying that the constant symbol zero is a numeral and if we have got something, call it θ which *is* a numeral, then $s(\theta)$ is a numeral, so the machine which just calculates whether something is the Gödel number of a numeral looks like this: first we plug in our number. Now zero has the Gödel number 1 and it is a sequence with just one member, so we get as the Gödel number of the sequence 0, the number $2^1 = 2$. So we first ask: is $n = 2$? If it is we are home and dry because that is just the number of the numeral 0. If the answer is NO then we factorize the number and ask if the exponents of the first two primes in the factorization are the number for s, which is 2, and the number for (, which is 6, and whether the last exponent is the Gödel number for the right-hand bracket, which is 7. Now if the answer is NO, then the thing cannot possibly be a numeral. If the answer is YES we remove the s, the (, and the) corresponding to the exponents 2, 6, and 7 and we recode as n^* and go back to the beginning. Then the numbers are getting smaller and smaller, so eventually we shall get the answer: NO, it is not a numeral, or: YES, it is a numeral. In either case the process ultimately stops.

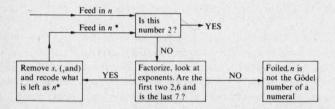

Now let us return to the definition of a formula (p. 12) and you will see another reason why we gave our definition of a formula in the particular way we did. In the system we have been considering, the only predicate letter that we have is $=$. So the basic formulae are just the meaningful equations. These will say that some (perhaps complicated) expression with plus, times, successor, etc. is equal to some other (perhaps complicated) expression. We disallow expressions like $0 = \rceil + \rceil$. The second clause of the definition says that, if A is a formula, then so is $\rceil A$ and also, if A and B are formulae, then so is $(A \ \& \ B)$. The third clause says that if A is a formula and v is a variable, then $\exists v A$ is a formula. (And in this set-up, a variable is an x with a string of little bar subscripts on it.)

Finally we have a clause that says that is all. So if we just look at this in terms of the actual formulae, in order to decide whether a given sequence of symbols is a formula or not, we first of all ask: Is it an equation? If the answer is YES, we are home and dry, the sequence of symbols is a formula. But if the answer is NO, then we can say: Does it start with the $\rceil$ sign? If it does, we knock this off and ask whether what is left is a formula. If the answer to this question is YES we are home and dry again. If the answer is NO, then we repeat until we have removed all the $\rceil$ signs. After this we may still have a formula, but not one starting with the $\rceil$ sign. Then we can try the next case: we ask if it is of the form $(A \ \& \ B)$, that is, a left bracket, a string of symbols, an ampersand, another string of symbols, and then a right bracket. If it is, we now ask whether A, B are formulae and if not we try the next case. Thus, finally, if that does not apply we ask whether it starts with an $\exists$ symbol. But in each case the string of symbols that we are looking at is getting shorter and shorter, so eventually the process is going to stop. Now we can do exactly the same as this for the Gödel numbers. What we can do is to construct a machine which is able to answer the question which I call Q: 'Is n the Gödel number of a formula?' Let me recall that by the Gödel number of a formula I mean a number of the form $2^p.3^q.5^r\ldots$ where the exponents are the Gödel numbers of the symbols in the formula.

In doing this as an example I am going to assume we already have at our disposal some machines for doing simpler things like deciding whether something is the Gödel number of an equation or of a variable. So what is the the lay-out of such a machine? First of all we take the number n and ask whether it is the Gödel number of an equation. If the answer is YES we can go back to sleep. If, however, the answer is NO we take further action; we ask whether n may be the Gödel number of an expression starting with a $\neg$ symbol. (See diagram.) To put this another way, does 2^{11} divide exactly into n and give an odd number? This is a perfectly straightforward numerical calculation, tedious but not impossible to perform. If the answer is YES then what we do is remove the $\neg$ symbol and recode, by which I mean: first remove the 2^{11} and then put the primes back again, starting with the smallest, 2, with the same sequence of exponents except for the first one (11). Of course for our previous example, $\neg x = 0$, what we have now will be the number of the formula $x = 0$. Now when we have the recoded number we go back to the beginning and start again. And when we feed in our recoded number (the number of $x = 0$) we shall get the answer YES.

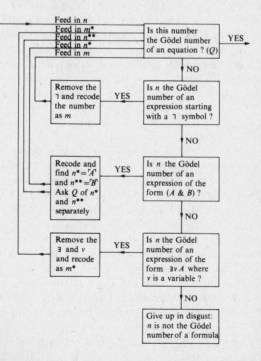

But there are other clauses of the definition of a formula and these we have not yet treated. So we have the possibility of getting a NO. If, when we started, we did not start with a $\neg$ symbol or an equation, it is possible that we had a conjunction. So we now ask (that is if we are still getting a NO): Is n the Gödel number of an expression of the form $(A \;\&\; B)$? Here A and B are not assumed to be formulae, just sequences of symbols.

If we do have an expression of this form, what we do is to take A and B and recode as n^* and n^{**} respectively. When I say recode now, I mean that n would have started off with a 2^6 and ended up with a prime to the seventh power. We take the exponents corresponding to the things in A and recode them as n^*. Similarly we get n^{**} from B and we shall write $\ulcorner A \urcorner$ for n^* and $\ulcorner B \urcorner$ for n^{**}. These will be smaller numbers than the one we started off with. And now we ask the question Q of n^* and of n^{**}. Is n^* the Gödel number of a formula? Is n^{**}? If they both come out as Gödel numbers of formulae then we give an answer to the whole question. If either comes out NO, then the machine says NO and we did not have a genuine formula. That finishes the case where we had the answer YES to the question: Did the formula look like $(A \;\&\; B)$?

If we did not have the answer YES, we still have one more trick to try. And we ask: is n the Gödel number of an expression which starts off with an $\exists$ symbol followed by a variable and then a string of symbols which we shall refer to as A? If the answer is YES, then remove the $\exists$ and the variable and recode A and send it back to the beginning. If the answer is NO, we give up in disgust. If you look at this machine carefully you will see that there are no loose ends. We have a YES or a NO coming out of it at each of the stages, and no matter what way we go, we will end up with a YES or NO. And even if we have to look at n^*, n^{**}, etc. these numbers are getting smaller and smaller so the process will always stop eventually.

Now you may say: What has all this business of being able to tell whether a certain number codes an expression got to do with Gödel's Incompleteness Theorems? It turns out that all the mechanical devices which you can have for checking whether numbers are the Gödel numbers of certain sorts of formulae or not are in a certain sense representable in arithmetic. These devices have corresponding formulae in arithmetic, so we can actually prove in arithmetic a result corresponding to what is happening in the real world.

The argument that I shall produce is essentially an analogue of the Liar Paradox, or at least is suggested by the Liar Paradox, although I shall present it in a slightly unfamiliar context. What we are going to do is produce a formula which says in effect 'I am not provable'. By virtue of the fact that we shall show that this formula is not provable, it will turn out that it is true. We shall have a formula in arithmetic which is true but not provable. Now from what I have already said we are almost at the stage of writing down in arithmetic some formulae corresponding to relations on ordinary (genuine if you like) natural numbers. I have in fact so far been talking about particular predicates being recursive. I have just gone to some lengths to show that the notion of being the Gödel number of a formula in our language of arithmetic is recursive: it is computable. We have a mechanical procedure for deciding whether a particular number is the Gödel number of a formula in this language or not.

I have not yet mentioned that we can code sequences of formulae too. We do it in just the same way that we coded sequences of symbols. We code the sequence of formulae, say ϕ_1, ϕ_2, ϕ_3 as $2^{\ulcorner\phi_1\urcorner}.3^{\ulcorner\phi_2\urcorner}.5^{\ulcorner\phi_3\urcorner}$ where $\ulcorner\phi_1\urcorner$ is the Gödel number of the formula ϕ_1 and so on. It hardly matters what coding procedure you follow, so long as it works in the sense that, given a number, you can first of all tell whether it is the code of a sequence or whether it is the code of a formula and secondly you can tell of just what that sequence consists.

Hence it now makes sense to say that n is the Gödel number of a sequence of formulae. Further, proofs are special sequences of formulae, so consider the relation $Pf(x, y)$. $Pf(x, y)$ is to be true only when x is the Gödel number of a sequence of formulae which constitutes a proof of the formula whose Gödel number is y. Now to check whether $Pf(x, y)$ holds between any two numbers is a mechanical (computable) procedure. You may recall the way we defined proof of a particular formula (*see* p. 13): a proof is a sequence of formulae (and we can check whether a thing is a formula, so we can check whether we have a sequence of formulae) such that each formula is either an axiom or else is obtained from earlier formulae in the sequence by the rule of inference. Now we can tell whether a formula is an instance of an axiom or not, because the axioms are given in a recognizable form. Also the rule is very simple: all that is involved is a routine 'chopping up' of the formulae given. This is a purely mechanical thing, so the predicate $Pf(x, y)$ is computable. Now I shall use recursive and computable as synonyms, so in other words $Pf(x, y)$ is recursive. All computable procedures can be performed by Turing machines, so we can actually construct a Turing machine which will do exactly what we want it to do here, that is, decide whether $Pf(x, y)$ holds.

Recursive predicates have a very special property: they are exactly the ones which are representable in the following sense. Let us write $\bar{n}$ for the numeral for n so $\bar{2}$ is $s(s(0))$. Suppose we have a relation between natural numbers $R(m, n)$; then I shall say that $R(m, n)$ is representable in arithmetic (and from now on I shall drop the phrase 'in arithmetic') if there is a formula of arithmetic $R(x, y)$ such that

(i) If $R(m, n)$ is true then $\vdash R(\bar{m}, \bar{n})$. (It makes no sense to say that $\vdash R(m, n)$ because m, n are not numerals.)

(ii) If $R(m, n)$ is false then $\vdash \neg R(\bar{m}, \bar{n})$.

(In this relation there are two arguments x, y; in fact there could be any number from zero upwards depending on the relation we are considering.) The interesting point is that there are relations that do not have this property. In fact this is the essence of Gödel's First Incompleteness Theorem. We find something which is true, but neither it nor its negation is provable.

It is very easy to see that some relations are representable. The relation of equality is representable by the formula $x = y$ because if m equals n then $\vdash \bar{m} = \bar{n}$ (i.e. $\vdash s \dots (m \text{ times}) \dots s0 = s \dots (n \text{ times}) \dots s0$). We can prove that $\vdash \bar{m} = \bar{n}$ is a theorem of our system from the axioms which say $\neg 0 = sx$, etc. The proofs are fairly elementary. Also we have that if m is not equal to n then $\vdash \neg \bar{m} = \bar{n}$.

Now one more example. It is fairly obvious that addition and multiplication are representable in the system because we included plus and times as symbols in our language. But division is also representable, because the relation 'm divides n' is represented by the formula $\exists z(x \times z = y)$. If m divides n then in arithmetic we can prove $\exists z(\bar{m} \times z = \bar{n})$ and if m does not divide n then we have $\vdash \neg \exists z(\bar{m} \times z = \bar{n})$. In fact one can show that every recursive relation is representable. We shall take just two special cases, one more complicated than the other.

Consider the relation $Pf(x, y)$ which holds only when x is the Gödel number of a proof of the formula whose Gödel number is y. Then there is a formula $Pf(x, y)$ in arithmetic such that if $Pf(m, n)$ is true, we can actually prove in arithmetic the formula $Pf(\bar{m}, \bar{n})$. And if m is not the Gödel number of a proof of the formula with Gödel number n (or indeed if n is not the Gödel number of a formula at all, which is quite possible), then we can prove $\neg Pf(\bar{m}, \bar{n})$. The reason we can do this is that this relation is recursive, and I already said that all recursive relations are representable. Let us observe that if we can prove $Pf(\bar{m}, \bar{n})$ then m really is the Gödel number of a proof of the formula with Gödel number n, because if this were not so, then we would have $\vdash \neg Pf(\bar{m}, \bar{n})$ and thus a contradiction. Now for the more complicated relation $Pf^+(x, y, z)$. Its representing formula gives us essentially the diagonalizing formula. (I put a superscript $^+$ on it, so that we shall not confuse it with the previous relation.) The relation is to hold only when y is the Gödel number of a formula with just one free variable and x is the Gödel number of a proof not of the formula whose Gödel number is y, but of a slight variant of it, namely the formula that results when the numeral for z (that is $\bar{z}$) has been substituted for the free variable in the formula. This again is a recursive predicate, because if you are given the three numbers x, y, z you can certainly decide whether y is the Gödel number of a formula, and if you write that formula out, then you can decide whether it has one free variable or not. If it does not, then the predicate is false. If it does, then we go to the next stage and we look at the number x. We check mechanically as to whether it is the Gödel number of a proof of the formula that we have just found when we have made a certain substitution in that formula, namely when we have put an appropriate string of s's (z of them) followed by 0 and appropriate brackets instead of the free variable in the formula that we are talking about.

If x is such a number then the relation is true and if it is not, then the relation does not hold. So it is obviously a mechanical procedure. But all mechanically checkable formulae are representable, so $Pf^+(x, y, z)$ is representable. I shall write $Pf^+(x, y, z)$ for the formula which represents it.

Now let us consider the formula which says that x is the Gödel number of a proof of the formula that you get when you put in for its free variable the numeral for its own Gödel number. And this is very similar to the diagonal argument from the first chapter where you are enumerating the real numbers and in the nth decimal you change the nth decimal place. Here we have got the yth formula and we are putting in $\bar{y}$ in the yth formula. The formula $\neg \exists x\, Pf^+(x, y, y)$ has one free variable, namely y. It has a Gödel number and let us call its Gödel number g. I call the formula itself G. Now let us substitute for the free variable y in G the numeral for G's Gödel number; that is substitute $\bar{g}$ for y. The free variable appears twice, so we have to put $\bar{g}$ in twice. What I want to show is that neither this formula nor its negation can be proved in arithmetic. Let us consider for a moment what this formula says. When I say: 'What this formula says' I really mean: let us go back to the relation which this formula corresponds to. Reading it informally it says that it is not the case that there is a number x such that x is the Gödel number of a proof of the formula with Gödel number g (which is just the formula $\neg \exists x\, Pf(x, y, y)$ when for the single free variable y in it we have substituted the numeral for the number g). In other words, there is no proof of this formula when we have put the numeral for g in place of the free variable y. But the resultant formula is exactly the formula G itself. So what the formula G says is: there is no proof of formula G. So it would be somewhat surprising if we could prove that formula, and indeed we cannot. This is what I shall now proceed to show.

Let us first of all suppose that we can prove G, in other words, let us suppose $\vdash \neg \exists x\, Pf^+(x, \bar{g}, \bar{g})$. If we can prove it, then there is a proof, and this proof has a certain Gödel number, say m. So m is the Gödel number of a proof of the formula which we get from the formula whose Gödel number is g by substituting the numeral $\bar{g}$ for its free variable. So $Pf^+(m, g, g)$ is true. But if it is true then, as $Pf^+(x, y, z)$ is represented by $Pf^+(x, y, z)$ in arithmetic, we can actually prove the corresponding formula $Pf^+(\bar{m}, \bar{g}, \bar{g})$. But then we prove $\exists x\, Pf^+(x, \bar{g}, \bar{g})$ by elementary predicate calculus. So as we are assuming that arithmetic is consistent, we have a contradiction. Thus our assumption was false and $\neg \exists x\, Pf^+(x, \bar{g}, \bar{g})$ is not provable.

Suppose on the other hand we could prove the negation of this formula. In other words, suppose that we could prove $\neg \neg \exists x\, Pf^+(x, \bar{g}, \bar{g})$. When we have two negations, they 'cancel' each other. So we can prove $\exists x\, Pf^+(x, \bar{y}, \bar{y})$. Now we have just shown that we cannot prove $\neg \exists x\, Pf^+(x, \bar{g}, \bar{g})$. If you cannot prove it, you cannot have any number which is the Gödel number of a proof. So 0 is not the Gödel number of a proof of G, 1 is not the Gödel number of a proof of G, and so on. Now what do these things say? They say that

$Pf^+(0, g, g)$ does not hold, and that $Pf^+(1, g, g)$ does not hold, and so on. These are all false.

If we have all of these things being false, then we can actually prove (because $Pf^+(x, y, y)$ is representable) the negation of $Pf^+(\bar{n}, \bar{g}, \bar{g})$ for each number n. If we can do this for every number n, it would be highly implausible to be able to prove $\neg\forall x\, \neg Pf^+(x, \bar{g}, \bar{g})$. (I shall say more about this later.) But what are we saying? We are saying that we cannot prove $\exists x\, Pf^+(x, \bar{g}, \bar{g})$. But we assumed that we could prove it, so that is impossible too. So we have got to the stage where neither the negation of this formula nor the formula itself is provable. As I remarked earlier, this formula says 'I am not provable', so indeed its interpretation is true: it is not provable. So this establishes Gödel's Incompleteness Theorem in two senses. The stronger sense is this one: that there is a formula such that neither it nor its negation is provable. The weaker sense is the corollary that there is a formula that is true but not provable in arithmetic.

I conclude this chapter with one or two remarks concerned with consistency in one shape or another. We have been assuming all along that arithmetic is consistent and we have very good reason for doing that. Namely that the ordinary arithmetic that we deal with in day-to-day life obviously satisfies the axioms that we have, so these axioms are true and we know the rules of proof preserve truth. So therefore all the things that we can prove are true, so we cannot have an inconsistency. However, we invoked a slightly subtler form of consistency in the above argument. From having $\vdash \neg Pf^+(\bar{n}, \bar{g}, \bar{g})$ for each number n we concluded we could not have $\vdash \neg \forall x\, \neg Pf^+(x, \bar{g}, \bar{g})$. So you might like to put in a remark at this stage of the proof that we are assuming (what is known in the trade as) ω-consistency. (Here ω just refers to the totality of the natural numbers.) We say that a formal system containing, for each natural number n, a notation (or numeral) $\bar{n}$ for that number, is ω-consistent if whenever we can prove $A(\bar{n})$ (where A is just some formula) for each natural number n, it is not the case that there is something else lurking about which enables us to get a proof of $\exists x\, \neg A(x)$. I think you will admit this is a very reasonable thing to assume. If you take this formula A just to say something simple like $x = x$ then it is clear that if a system is ω-consistent, then it is consistent in the usual sense. But the converse is not true. We can have systems which are consistent but which are not ω-consistent. I only put this in for thoroughness, as I do the next remark also. It is possible to soup up the definition of the formula $Pf^+(x, y, z)$ so that the condition of ω-consistency can be dispensed with. This result is due to Barkley Rosser.

We considered a formula which says it is not the case that $\exists x\, Pf^+(x, g, g)$ and the modification that Rosser produced was to say if x is the Gödel number of the ghastly formula $\neg\exists x\, Pf^+(x, \bar{g}, \bar{g})$ then there exists a smaller number which is a proof of the negation of that very same formula. And as you can imagine the explanation of the formula gets a little bit more tedious, but the argument goes through in a very similar vein to the one I have given.

Now I want to make one final remark. I want to talk about Gödel's Second Incompleteness Theorem. This requires not only talking about provability in the system, but, shall we say, the limits of provability in the system. Gödel's Second Incompleteness Theorem is the following: the consistency of arithmetic is not provable in arithmetic. And by arithmetic I mean formal arithmetic as I have always done in the course of this chapter. If you take $0 = 1$, then that gives you a contradiction with the axiom which says that 0 is not the successor of any number. So if we prove this formula, we shall have an inconsistent system, and if we had an inconsistent system we could prove this formula. So not being able to prove this formula is equivalent to having consistency. Now this formula has a certain number. Let this number be k. Then let us consider the formula $\forall x \,\neg \mathrm{Pf}(x, \bar{k})$. What does it say? Its interpretation is that for all x, it is not the case that x is the Gödel number of a proof of the formula with Gödel number k. In other words, it says that there is no proof of the formula $0 = \bar{1}$. This is how I shall read $\forall x \,\neg \mathrm{Pf}(x, \bar{k})$. So let Consis be a name for this formula. Now I shall not say that what we do next is easy to show, because it is quite difficult. But it is more a matter of routine than anything else. What we can actually do is take all the argument which I have been presenting and formalize that. We formalize our previous argument in formal arithmetic and we can prove the formula $(Consis \rightarrow G)$, that is to say we have $\vdash (Consis \rightarrow G)$. (Recall that G is the formula $\neg \exists x \, \mathrm{Pf}^+(x, \bar{g}, \bar{g})$.) We take all our previous argument and put all that too within arithmetic. Everything was done in a very finitistic manner and so we can get that $(Consis \rightarrow G)$ is provable. But now if we could prove in arithmetic the formula which asserts the consistency of arithmetic, if we had $\vdash Consis$, then (by modus ponens) we could prove our formula G. We have shown that this is impossible. So it must be false that we can prove the consistency of arithmetic within arithmetic.

So that is what Gödel's Second Incompleteness Theorem says. Arithmetic is not sufficient to prove its own consistency. And indeed any extension of arithmetic has this same deficiency, so there is no hope of proving the consistency of all mathematics from a formal system for all mathematics. In order to get consistency we have to go into a bigger system and this system requires more than the ordinary mathematical induction which we introduced as an axiom scheme; what is required is called transfinite induction (up to a not very large ordinal number). We can give a consistency proof for arithmetic as a direct result of transfinite induction up to the first epsilon number, as it is called.

In the next chapter, bigger systems, systems of set theory, will be considered. However, set theory contains arithmetic, and so has the same deficiency. Consistency problems arise there and you will see that we are very much concerned with relative consistency rather than absolute consistency for the reason that we cannot ever hope for absolute consistency because of Gödel's Incompleteness Theorems.

6
Set Theory

IN this last chapter, I am going to discuss set theory. Probably you all have some sort of picture of your own, some rough idea of what a set is: a sort of collection of things. I shall start by explaining a number of axioms for sets; the axioms that most mathematicians agree are true about sets.

I shall do this very informally, though when you come to the crux of the matter, Axiomatic Set Theory is really a formal theory of the type that has been described earlier in this book. It is based on the predicate calculus with equality, with the one additional predicate symbol $\in$ for set membership. I shall write $x \in y$, and read it as 'x is a member of y'. All the axioms which I shall give can be written out appropriately in this language. The particular formal system that results is called ZF, for Zermelo–Fraenkel set theory.

What sort of axioms do we believe are true? What sort of picture do we have of sets first of all? I suppose that the best we can say to start with is this: sets are collections of mathematical objects. And to start with I cannot be very much more precise. The first principle that this leads me to suspect may be true is the following. Suppose that I have any property that applies to sets. Then there is the set of all things with this property. Thus we have the principle:
(P) For any property $\psi(x)$, there is a set y such that

$$x \in y \leftrightarrow \psi(x).$$

Now in fact this principle is not correct, for from it we can derive Russell's Paradox (found by Russell in about 1902). Apply (P) to the property $x \notin x$: there must be a set R such that

$$x \in R \leftrightarrow x \notin x.$$

So R is the set of all sets which are not members of themselves. If you come to this set R, and ask it the question: 'Do you belong to R?' the set is rather embarrassed and does not know how to answer, because R belongs to R

exactly when it does not:

$$R \in R \leftrightarrow R \notin R,$$

as you may check easily.

The fact that Russell's paradox does occur is the reason why we need to bother with a theory of sets at all. If the principle (P) were true, then there would be little else that we need say. But the fact that Russell could derive his contradiction from this first innocent sort of an axiom shows us that we have to look rather carefully at our intuitive view of a set. It is not quite as obvious as it first seemed to us.

To get a slightly better idea of what sort of things should be sets, one must analyse where the principle (P) goes wrong. Why is the Russell collection not a set? It turns out that it is not a set because there are too many things that want to be in it. R is just too big a collection of things to be a set. So in the light of this experience we shall modify slightly our intuitive view of a set to the following: a set is a not-too-big collection of things.

And the axioms that I shall write down are axioms that try to capture this view.

For the first one, let us dispose of equals, that is, let us get the connection between $\in$ and $=$ in our formal theory. It follows at once from the axioms for equality that if two sets are equal, then they have the same members. Now the only thing that I have in my hands for distinguishing between sets is the membership relation. So if two sets have exactly the same members, then I cannot distinguish between them, in other words, they are equal. And this is what the first axiom is going to say. This is the Axiom of Extensionality,

$$\forall z(z \in x \leftrightarrow z \in y) \rightarrow x = y.$$

In the next axiom let us rescue what we can from the set-building principle (P) that we had to start with. Certainly there ought not to be any harm if instead of trying to collect together all the sets that have some property, we start with a given set and collect together just the members of that set which have the property in question. This seems reasonable because, if anything, this is cutting down the size of a set that we have already got, and that should not come to something that is too big! This leads to the Comprehension Schema. When I try to write it in the formal system it has to be a schema. For each property I have to say that a particular set exists. Now what do I mean by property? If you think of a formal system, there is no question of what a property is. It is really a formula with one free variable. So the Comprehension Schema is the following:

Given any set a, and given any formula $\psi(x)$ in which x is a free variable (there may be other free variables as well, they do not bother us), then there is a set b such that the members of b are exactly those members of a which have the property ψ, that is

$$x \in b \leftrightarrow x \in a \ \& \ \psi(x).$$

It follows from the Axiom of Extensionality that this set b is the unique set with this property. I shall write

$$\{x \in a : \psi(x)\} \quad \text{for } b.$$

This schema captures as much as we can of the first intuitive principle which was not quite good enough in itself.

What else can we construct? If we have got two sets, surely we can form a set that has exactly these two sets as members. This is the Pair Set Axiom: given two sets x, y there is another set which has exactly x and y as members. I write this new set as $\{x, y\}$.

Now the next thing, the Power Set Axiom. If I have a set, then I can think of all possible subsets of this set. It is probably going to be a larger collection, but not so terribly much larger. It is reasonable to think of this as giving us back a set. This leads to the Power Set Axiom: if I am given any set x, then the power set of x is again a set. The power set of x is the collection of those sets y which are subsets of x: $Px = \{y : y \subseteq x\}$. (I write $y \subseteq x$ to mean that y is a subset of x: all members of y are members of x.)

The next axiom is concerned with the union set of a set x. Take a set x to start with. We can think of x as being represented by a circle, with its members inside: But what are these members? They must themselves be sets, and so

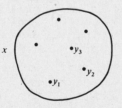

they themselves have members, so I can represent them in the same way (I have drawn only a few of them).

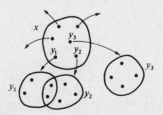

I want to assert that there is a set having in it exactly the things in one of $y_1, y_2, y_3, \ldots$. This set is called the union set (or the sum set) of x. I write it as $\bigcup x$. Thus the members of $\bigcup x$ are exactly the members of the members of x,

$$\bigcup x = \{z : \exists y (y \in x \ \& \ z \in y)\}.$$

The Union Set Axiom says that for any set x, indeed $\bigcup x$ exists, and is again a set. Notice what happens if I take two sets a, b, and form $\bigcup \{a, b\}$. It is easy to see that $\bigcup \{a, b\}$ is just the union $a \cup b$ of the two sets a and b. So this axiom has the consequence that the union of two sets a and b is always a set again.

One of the reasons for inventing set theory is to enable us to talk about a collection of infinitely many objects. So we had better have an axiom to say that there are sets of this sort. This is the Axiom of Infinity: there is an infinite set.

Now that we know that there is a set (indeed, an infinite one), we can prove that there is a set which has no members at all. Let ω, say, be an infinite set, as provided by the Axiom of Infinity. Consider the following set $\emptyset$, which the Comprehension Schema asserts to exist:

$$\emptyset = \{x \in \omega : x \neq x\}.$$

Since $x = x$ for all sets x, it follows that $\emptyset$ can have no members. In fact, $\emptyset$ is the unique set with this property. We call $\emptyset$ the empty set, or the null set.

The next axiom that I want to go on to is not quite the same building-up sort of axiom that we have had so far. Consider the picture that I had when I was talking about the union of a set x. I started off with a set x and said: think of the members. If I think of the members, I can draw them out. These sets have members themselves. Let me take one of these. This is equally well a set, so I can think of it having members too. Now if I take one of the members of this set I can do the same thing again. When is this going to stop? Can I

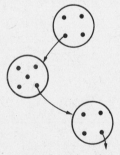

just keep on doing this forever? My intuitive view of a set certainly does not allow us to go infinitely far downwards like that. So let me make an axiom that says: whenever I try to chase down a chain of members, it must stop at some finite stage. You can think of it this way: we have a string of sets $x_1, x_2, x_3, \ldots$ where each is a member of the preceding one, that is $\ldots \in x_3 \in x_2 \in x_1$.

This we shall call a descending membership chain. Then the axiom, the Axiom of Foundation (or of Regularity) is this: any descending membership chain is finite.

Using this axiom, we can show that never is it the case that a set x is a member of itself. For suppose, on the contrary, that you had a set x such that $x \in x$. Think of an infinite string of sets, all of them being x, thus $x, x, x, \ldots$. Then we get the infinite descending membership chain $\ldots \in x \in x \in x$. But the axiom says that there are no such chains, and we have derived a contradiction. This shows that $x \notin x$ for all sets x.

The next axiom that I want to go on to is the one that really captures the fact that I tried to suggest before, that any not-too-large collection of sets is itself a set. It is going to make quite precise what not-too-large means. It is going to say exactly that any set-sized collection of sets is a set. Now how can we phrase this in set-theoretic language? We want to start off with a given set a, and make another set correspond to each element of a. Then we shall assert that the collection of all the corresponding things is again a set. I can talk about a correspondence by means of a formula $\psi(u, v)$ of the formal language, with (at least) the variables u and v free. ψ is to make v correspond to u, and since I want each u to have just one set corresponding to it, ψ must have the function-like property:

$$\psi(u, v) \,\&\, \psi(u, w) \rightarrow v = w.$$

Thus the axiom will again be a schema: there will be one axiom for each function-like formula. So the schema is this: given a function-like formula $\psi(u, v)$ and a set a, there is a set b such that the members of b are exactly those sets v which correspond under ψ to some set u belonging to a, that is

$$\exists b(v \in b \leftrightarrow \exists u(u \in a \,\&\, \psi(u, v))).$$

This is called the Replacement Schema: it replaces these sets u in a by the corresponding sets v. It is this schema which really captures the fact that set-sized collections of sets give us back sets. And in fact once we have this, we can throw out the Comprehension Schema because it comes as a consequence of the Replacement Schema. But I am not the least bit worried about the fact that some of the things I am listing can be deduced from other ones.

Now there is only one more axiom that I need to complete the list of axioms of ZF. Let us keep it separate from the rest, for, whereas all mathematicians working in conventional mathematics accept the others as being true of the sets with which they work, some people are a little more dubious about the one that I am about to mention. This one is the Axiom of Choice, Russell's Multiplicative Axiom. This one says that if I have a set of non-empty sets, then I can choose one member from each set in this collection and put the chosen elements together into a set.

Before I discuss this further axiom I shall put it in another form which is somewhat more mathematical. Let me observe that from the axioms that we have at this stage I can say all the things that one wants to say in ordinary mathematics. I can talk about ordered pairs, for instance. An ordered pair $\langle x, y \rangle$ is a set which is uniquely determined by its first component x and its second component y, in the sense that

$$\langle x, y \rangle = \langle u, v \rangle \leftrightarrow x = u \ \& \ y = v.$$

It turns out that we can define a set $\langle x, y \rangle$ which has this property. (You may care to meditate upon the set $\{\{x\}, \{x, y\}\}$.)

Now with this concept of ordered pair I can say precisely when a set is going to be a function, a mathematical function. A function makes the members of one set correspond to the members of another set. Now I can express this precisely: you make the correspondence between the x's of one set and the y's of another by having in the function the ordered pairs $\langle x, y \rangle$. And the functions can be just the collection of all such ordered pairs, the x's and the y's that correspond. So a function is a set of ordered pairs, with the right function property.

More relevant to the Axiom of Choice, let me now say what a choice function is. If x is a set, then a choice function for x is a function f which has the property that whenever I have a non-empty element of x, say y, then f makes a member of y correspond to y. I can write it this way: if $y \in x$ and $y \neq \varnothing$, then $f(y) \in y$. So, if you like, the f chooses for us a member of each of the y's. Such a thing I call a choice function for x: it chooses something from all the non-empty sets in x. Now I can express the Axiom of Choice rather more precisely: every set has a choice function.

I am going to return to the Axiom of Choice later. For the moment, I think it is appropriate to point out why people who accept the rest of the axioms are sometimes a little more dubious about the Axiom of Choice. To me it is true. It is clear that if I have a collection of non-empty sets, certainly I can choose a member from each. And in fact people working in set theory at the beginning thought it was so true they did not even notice they were using it. However, when people analysed their arguments more closely later, they realized that there was a principle involved in being able to make a whole lot of choices. Now the reason some people are a little more dubious about this than the others is this: let us go back and look at the other axioms and the things that they say are sets. These axioms give us a concrete construction for sets. Given a set x, they say that there is another set, and they give the recipe for producing it. Now the Axiom of Choice is not quite as definite as that. All it says is: given a set there is a choice function, but it does not say how to get the choice function. And some people are a little more dubious of the Axiom of Choice for just that reason. In fact the people who deny the Axiom of Choice are really doing themselves a disservice when they are thinking about the theorems they can prove, in the following sense: one can show that, if one can derive a contradiction using the Axiom of Choice, then one in fact could have derived a contradiction without using the Axiom of Choice. So there is no harm in using the Axiom of Choice if you are just worried about deriving a contradiction or not. I shall try and outline how this happens later.

Now I think it would be rather fun to use the axioms that we have listed and first of all recover those definitely mathematical objects $0, 1, 2, 3, 4, \ldots$. And then I would like to extend them out beyond infinity and go into the transfinite with these numbers. And in fact I am going to extend them in two ways—two different ways concerned with the two different ways in which one uses numbers anyway.

The first things that I want to consider are the ordinal numbers. One can think of the ordinal numbers as counting off places in a queue. Suppose that you are going to the theatre and there are a number of ticket windows and queues of people at the ticket windows. You wish to know which is the shortest queue to get on. How do you decide? You number off the places in each queue, so there is a first place, second place, and third place in this one, a first, second, third, fourth, and fifth in the next one, and so on. In other words you compare each queue with a standard that you hold in your head. And you pick as the queue to get on the one that goes the shortest distance down your standard. Consider now how sets would find this. Think of a whole lot of sets lining up at the ticket offices outside a picture theatre. And if one poor set is a bit late, he comes rushing down and wants to know which is the best queue to get on, that is, which is the shortest queue. Now he would want to do something the same as we did above. He would want to start numbering off each queue. So he looks at a queue. He sees a first place, a second place, a third place, $\ldots$, and so on. However, there are so many sets that even when he has numbered first, second, third, $\ldots$, there are still a whole lot more sets in the queue. Let us call the place after all these the ω-th place. Then there is the place after that: the $\omega + 1$st place. Then there would be the $\omega + 2$nd place. And so on: it would keep on going. Now what can we construct as a standard for this set to carry in his head? He must be able to number off the places in a queue that goes on just like the one we have been talking about. If it is a standard that he can carry in his head, it must be the simplest possible sort of thing we can imagine. So I want to construct the standard queue that every set can carry in his head and can compare other ones against. Now let us think, what are the easiest things to use? He wants to know when one thing comes before another. What is the simplest relation that two sets can have? Surely the membership relation. Let us then try and build up a standard queue where things will be related by membership. So the standard will be a queue ordered by membership. A particular y will belong to the next thing after y in the queue. It is rather convenient, in fact, if we are able to determine all the predecessors of x in the standard queue, just by looking at x without having to look at all the things before it. So let us ask that all the predecessors should also belong to x. We have not only the thing immediately before, but everything before, belonging to x.

What should be our first thing in the standard queue? We should start with the simplest set there is, in other words, with the empty set. On the suggestion above, what should come next after the empty set? Just all its predecessors. And is $\varnothing$ not its only predecesser? So it should be the set whose only member is $\varnothing$, in other words $\{\varnothing\}$. (Note that there is a big difference between these two objects, for $\varnothing$ has no members at all, but $\{\varnothing\}$ has exactly one member: it has $\varnothing$ as its member.) Now what is going to be next after $\{\varnothing\}$? Because it has to have all earlier things in it, it has to have $\varnothing$ and $\{\varnothing\}$. So this set is $\{\varnothing, \{\varnothing\}\}$. The one that comes after $\{\varnothing, \{\varnothing\}\}$ should have all the earlier things in it. So it is $\{\varnothing, \{\varnothing\}, \{\varnothing, \{\varnothing\}\}\}$. And the next one is: $\{\varnothing, \{\varnothing\}, \{\varnothing, \{\varnothing\}\}, \{\varnothing, \{\varnothing\}, \{\varnothing, \{\varnothing\}\}\}\}$. We keep building them up like that. As you can see, it is getting a little bit tedious to write these down.

So I want a shorthand for them. What should I call them? If you look at them, there is an obvious suggestion that comes out. The first has no members, the second one has one member, the third has two members, and so on. So let us actually define these as zero, one, two, etc. Thus

$$0 = \varnothing$$
$$1 = \{\varnothing\} = \{0\}$$
$$2 = \{\varnothing, \{\varnothing\}\} = \{0, 1\}$$
$$3 = \{\varnothing, \{\varnothing\}, \{\varnothing, \{\varnothing\}\}\} = \{0, 1, 2\},$$

.
.
.

Now look what happens. 1 is the set where the only member is zero. 2 is the set whose members are zero and 1, and so on. Thus we have in general that the integer n is the set of all the smaller integers: $n = \{0, 1, 2, 3, \ldots, n - 1\}$. Thus we can define the natural numbers to be particular sets; in fact we can take for the natural number n the set of all the smaller natural numbers.

We must return to our standard queue again. The natural numbers are just the beginning of it. So the standard starts $0, 1, 2, 3, \ldots$. What is going to happen at the ω-th place? After $0, 1, 2, 3, 4, \ldots$ what should follow? We said it was to be the set of all the earlier things. So the thing that comes at the ω-th place, let me call it in fact ω, should just be the set that has in it $0, 1, 2, 3, \ldots$. So $\omega = \{0, 1, 2, \ldots\}$. Now what happens at the $\omega + 1$st place? It has to have in it all the things before. So the $\omega + 1$st thing, which we might as well call $\omega + 1$, should have in it $0, 1, 2, \ldots$ and ω too. So $\omega + 1 = \{0, 1, 2, \ldots, \omega\}$. Let us go just one step further at this stage. What is $\omega + 2$? We shall have $\omega + 2 = \{0, 1, 2, 3, \ldots, \omega, \omega + 1\}$. And in this way one can build up a standard sort of queue that a set can carry in its head and against which it can number off any other one.

The sets that can be constructed in this way, that fall into the standard queue, are what are called the *ordinal numbers*. I shall use $\alpha, \beta, \gamma, \ldots$ in future for ordinal numbers. And from the way we have constructed them, it turns out that the ordinal number α is the set that has in it as members exactly these ordinal numbers β which are before α in the standard queue. So we have: α is an ordinal exactly when $\alpha = \{\beta : \beta$ is a smaller ordinal than $\alpha\}$. The ordinal numbers that we get in this way are going to be very important for all that we do from now on.

Now I start considering another way of extending the natural numbers into the transfinite. I shall deal with the *cardinal numbers*. Each set has associated with it exactly one cardinal number, and this tells us the size of the set, how many elements it has.

When should we say that two sets have the same size? Think what it means in the finite case first of all: if I have a bag of apples and a bag of oranges, how do I decide if there is the same number of apples as of oranges? What I do is to pair off oranges with apples. I take an orange from the one bag and an apple from the other and put those two together. Then I repeat the process, and if I can exactly pair off *all* the oranges with *all* the apples, then I say there are as many oranges as there are apples.

We can use a similar concept to decide when two sets have the same size. Let us say that two sets have the same size exactly when all the members of one can be paired off with all the members of the other. More precisely, this means there is a function which will establish the correspondence. For sets x and y, write $x \approx y$ when this holds, so:

$x \approx y \leftrightarrow$ There is a function which pairs off all the elements of x with all the elements of y.

In a similar way, I can say when x is of size less than or equal to that of y: all the members of x can be paired off with some (but possibly not all) of the members of y. Thus

$x \leqslant y \leftrightarrow$ There is a function which pairs off all elements of x with some of the elements of y.

If some elements of y are always left over no matter how we try to pair off x and y, in other words if $x \leqslant y$ but $\neg(x \approx y)$, then I write $x \prec y$.

For example, if n is any finite ordinal, then $n \prec \omega$.

It is of interest to note, however, that

$$\omega \approx \omega + 1.$$

Remember that $\omega = \{0, 1, 2, \ldots\}$, and that $\omega + 1 = \{0, 1, 2, \ldots, \omega\}$. Now to show $\omega \approx \omega + 1$ I have to show that there is a function that pairs off the elements in $\omega + 1$ with those in ω. Take this one: it says the last shall be first, and all the rest move down one. Diagrammatically

If this is to be a useful concept of size we must certainly be able to compare any two sets given to us. Thus we need to know, for any sets x, y that

$$x \leqslant y \quad \text{or} \quad y \leqslant x.$$

This is indeed provable from our axioms. In fact it turns out to be equivalent to the Axiom of Choice. This just gives more evidence for the truth of the Axiom of Choice.

Does this concept of size give all the infinite sets the same size? It would not be of much use if it did. Fortunately, however, there are lots of sizes of infinite sets. For take any set a, then I can show you a set that is definitely of larger size—the power set of a. This is a result known to Cantor (see Chapter 1).

CANTOR'S THEOREM: For any set a, $a \prec Pa$.

Proof: It is always the case that $a \leqslant Pa$, for I can pair off each $x \in a$ with $\{x\}$, and certainly $\{x\} \subseteq a$, so that $\{x\} \in Pa$. So this pairs off each element of a with some element of Pa, so showing $a \leqslant Pa$.

Now suppose $a \approx Pa$. So there is a function f that pairs off all the members of a with all the subsets of a. Take $x \in a$, then $f(x) \subseteq a$, so we may ask if $x \in f(x)$. Consider the set b,

$$b = \{x \in a : x \notin f(x)\}.$$

Since $b \subseteq a$ there is a unique $y \in a$ such that $b = f(y)$. Let us ask whether or not $y \in f(y)$.

 (i) If $y \in f(y)$, then $y \in b$, and so $y \notin f(y)$ by the definition of b.
(ii) On the other hand, if $y \notin f(y)$, then $y \notin b$. But if $y \notin f(y)$ then by the definition of b, $y \in b$.

Thus, in either case, we reach a contradiction. This was derived from the supposition $a \approx Pa$, so we must conclude $a \not\approx Pa$.

Hence $a \leqslant Pa$ & $a \not\approx Pa$, that is $a \prec Pa$.

So this establishes that given any set, in particular an infinite one, I can give you one of larger size by taking its power set. Since I can repeat the process again and again, there are thus lots of sizes of infinite sets.

Reconsider the ordinal numbers: it turns out there is an ordinal of every possible size. (This statement also is equivalent to the Axiom of Choice.) Knowing that there are ordinals of every possible size, we can ask which ordinals are the first of any particular size. The answer is this: first there are the finite ordinals. There is exactly one of these for each size. 0 is the size of a set with no elements. Since $1 = \{0\}$, and so has (intuitively) one element, exactly the singleton sets can be paired off with 1. Likewise, just the two-element sets can be paired off with $2 = \{0, 1\}$, and similarly for the other finite ordinals.

What of the infinite sizes? The first infinite ordinal is certainly ω. So ω is a set of the smallest possible infinite size. (Such sets are called *denumerable*.) Now when am I going to find an ordinal of larger size? I know they exist, but which will be the smallest of them? It is not $\omega + 1$, because I showed $\omega \approx \omega + 1$. Neither is it $\omega + 2$, because the function saying 'the last two shall be the first two and the others move down two' shows $\omega \approx \omega + 2$. Likewise it is not $\omega + 3$ and so on. It is not even $\omega + \omega$, for the function mapping the first ω elements in $\omega + \omega$ onto the even elements of ω and the second ω elements in $\omega + \omega$ onto the odd elements of ω shows $\omega \approx \omega + \omega$.

Diagrammatically

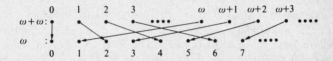

However, eventually we come to a bigger ordinal. Call it $\aleph_1$. ($\aleph$ is 'aleph' from the Hebrew alphabet.) Thus $\aleph_1$ is the first uncountable ordinal. The first ordinal bigger than $\aleph_1$ (we are guaranteed there are bigger ones) is called $\aleph_2$, the first after this $\aleph_3$, and so on. Since ω is the smallest infinite size in this context we call it $\aleph_0$.

Now we can define the *cardinal numbers*. They are just those ordinal numbers which are the smallest of their size. These we have seen are the finite ordinals and the sequence of alephs. Then any set is the same size as exactly one cardinal number, for it is the same size as some ordinal and there is only one smallest ordinal of that size. That unique cardinal number which is the same size as a given set x is called the *cardinality of x*, written $|x|$ (or sometimes $\bar{x}$). Thus $|x|$ *is* the size of x. From the definition of cardinality, as we should expect, it follows that

$$|x| = |y| \leftrightarrow x \approx y,$$

$$|x| \leqslant |y| \leftrightarrow x \preccurlyeq y.$$

Reconsider Cantor's Theorem. Applied to the set ω, it says

$$|\omega| < |P\omega|.$$

Let us write $2^{|x|}$ for $|Px|$, for any set x. Then Cantor's Theorem tells us:

$$\aleph_0 < 2^{\aleph_0}.$$

Now $2^{\aleph_0}$ is some infinite cardinal; it is bigger than $\aleph_0$. Which cardinal is it? $\aleph_1$? $\aleph_2$? ... Unfortunately set theory does not tell us. The axioms of set theory leave the matter undecided. The conjecture that $2^{\aleph_0}$ is as small as it can be, namely $\aleph_1$, is the *Continuum Hypothesis* (CH): $2^{\aleph_0} = \aleph_1$. (Cantor, in 1878, was the first to make this conjecture.) It is quite easy to show that $P\omega$ and the set of real numbers are of the same size, so the Continuum Hypothesis is a suggestion as to how many real numbers there are.

For *any* infinite cardinal $\aleph_\alpha$, we can ask which aleph $2^{\aleph_\alpha}$ is. Cantor's Theorem tells us that it has to be larger than $\aleph_\alpha$. The suggestion that it is as small as possible, (GCH) $2^{\aleph_\alpha} = \aleph_{\alpha+1}$ is the *Generalized Continuum Hypothesis*. The axioms of set theory do not settle its truth either.

Before we leave the cardinal numbers, let me emphasize how to test whether a particular ordinal number α is in fact a cardinal number. α is a cardinal exactly when there is no function available which will pair off (the members of) α with (the members of) any smaller ordinal. This brings about the following situation.

Suppose we have a model $\mathcal{A} = \langle A, \varepsilon \rangle$ for the theory ZF (where ε is the interpretation of $\in$), and a smaller model $\mathcal{B} = \langle B, \varepsilon' \rangle$ where B is a subset of A and ε' is just ε cut down from A to B. Then some of the ordinals that $\mathcal{B}$ sees as cardinals may very well not be cardinals in $\mathcal{A}$. For inside $\mathcal{B}$ we may not have available any appropriate 'pairing off' function, yet such a function may exist in the larger model $\mathcal{A}$. In particular, the ordinal that $\mathcal{B}$ thinks is the cardinal $\aleph_1$ may be just a denumerable ordinal in $\mathcal{A}$ and not a cardinal at all. (This is how the Skolem paradox was resolved in Chapter 3: the denumerable model there is $\mathcal{B}$, and $\mathcal{A}$ is 'the whole universe'.) This idea of how things happen *inside* a model will be most important in the final sections on forcing, where we shall consider denumerable models exclusively.

Consistency of Axiom of Choice

Now I want to show you that you cannot get a contradiction using the Axiom of Choice, if you cannot get one just using the other axioms. You will realize that this is merely another way of saying that the Axiom of Choice is consistent with the other axioms of set theory. You remember then that to show the Axiom of Choice is consistent with the other axioms, all we need do is find a realization (that is, a model) of the other axioms in which the Axiom of Choice is also true. The model that I am going to describe was first found by Gödel (in 1938), the model of 'constructible sets'.

The language of set theory has in it just the two predicate symbols $\in$ and $=$, so our model must give an interpretation for both these. It will be a normal model, so $=$ will be interpreted as true $=$. We shall also interpret $\in$ as $\in$, among the elements of the universe of the model. So it will be a model of the form $\langle L, \in \rangle$ (not mentioning $=$, by our convention for normal models). The members of the universe L are called *constructible sets*.

Since we are concerned about the consistency of the Axiom of Choice, I shall be careful not to use it in building up L. Nonetheless, it will turn out that this axiom is true in $\langle L, \in \rangle$, as well as all the other axioms.

Let me try to motivate the definition of a constructible set. Suppose you are given some particular set A, and can see nothing except the members of A. Think of yourself as shut up in a room that is A, able to see nothing but the elements of A. What sets can you talk about? In other words, what subsets of A can you name? We usually construct sets from the axiom schema of comprehension, so how would we proceed from this in A? The schema says, given any formula $\psi(v, v_1, \ldots, v_n)$ (here I have mentioned all the free variables in ψ), for any sets $x_1, \ldots, x_n$ we can find the set $\{x \in A : \psi(x, x_1, \ldots, x_n)\}$. Now consider our plight in A. There is a restriction on the sets that we can find to put in for the free variables of ψ: they must be members $a_1, \ldots, a_n$ of A. But this is not all: if ψ were to mention say Px, it may well be the case that x has subsets which are not members of A, and so we cannot talk about Px from inside A. Thus from within A we could not use $\{x \in A : \psi(x, a_1, \ldots, a_n)\}$. However, what we can do is talk about those x which, as far as A can see, have the property ψ that is, those x for which $\langle A, \in \rangle \vDash \psi[x, a_1, \ldots, a_n]$. So we can certainly talk about

$$\{x \in A : \langle A, \in \rangle \vDash \psi[x, a_1, \ldots, a_n]\},$$

where $a_1, \ldots, a_n \in A$. Sets of this form are said to be *definable in A*. Let us write $\mathrm{Def}(A)$ for the set of all sets definable in A, so formally
$\mathrm{Def}(A) = \{y : \text{for some formula } \psi(v, v_1, \ldots, v_n) \text{ and some } a_1, \ldots, a_n \in A, y = \{x \in A : \langle A, \in \rangle \vDash \psi[x, a_1, \ldots, a_n]\}\}$.

Let us have an example of a set definable in A. If $a, b \in A$, then I claim that $\{a, b\}$ is definable in A. Let $\psi(v, v_1, v_2)$ be the formula

$$v = v_1 \lor v = v_2.$$

Then it is clear that $\langle A, \in \rangle \vDash \psi[x, a, b]$ if and only if $x = a$ or $x = b$. So we have that

$$\{x \in A : \langle A, \in \rangle \vDash \psi[x, a, b]\} = \{a, b\}.$$

Hence $\{a, b\} \in \mathrm{Def}(A)$.

With the concept of 'definable in ...' we can get to Gödel's constructible sets. We start from the empty set and repeat again and again the operation of taking the set of sets definable in what we already have. To put this a little more precisely, I must first say what a limit ordinal is. A limit ordinal is an ordinal which has no ordinal immediately before it—one like ω, or $\omega + \omega$. On the other hand, ones like $\omega + 1$, or $\omega + \omega + 23$ are not limit ordinals. To define the constructible sets, think of the line of all ordinals. As we pass

each ordinal α we assign to it a set M_α, in the following way. First, $M_0 = \varnothing$. At the ordinal $\alpha + 1$ which is next after α, we put for $M_{\alpha+1}$ the immediately preceding M_α together with the set of all sets definable in this immediately preceding M_α, so $M_{\alpha+1} = M_\alpha \cup \mathrm{Def}(M_\alpha)$. When we come to a limit ordinal β, this recipe is meaningless (for there is no *immediately* preceding M_α). Here we collect together everything that we have obtained so far:
$M_\beta = \bigcup \{M_\alpha : \alpha < \beta\}$. Now finally, a set x is said to be *constructible* if x is a member of one of the M_α.

I write $L(x)$ for this, so

$$L(x) \leftrightarrow \exists\alpha \ (\alpha \text{ is an ordinal } \& \ x \in M_\alpha).$$

A word of warning: as we defined a model earlier, the universe of the model was a set. Now it turns out that L is not a set: it is too large to be one. However, it is possible to modify slightly the earlier definition, so that it still works for L. The main thing is to avoid writing $L \in x$, for L cannot be a member of any set. Once this is done, then $\langle L, \in \rangle$ does indeed give us a model for set theory. I shall not go into the proof, for parts of it are very difficult.

I shall content myself with showing why the Pair Set Axiom is satisfied. Take two constructible sets x, y. Then I need to find a constructible set which L believes is their pair set. Clearly it is enough to show that their true pair set is constructible. Now we have ordinals α, β for which $x \in M_\alpha$ and $y \in M_\beta$, and we may as well assume that $\alpha \leqslant \beta$. Then it turns out that $M_\alpha \subseteq M_\beta$, and so both x, y are in M_β. But as I showed earlier, this means that $\{x, y\} \in \mathrm{Def}(M_\beta)$. Since $\mathrm{Def}(M_\beta) \subseteq M_{\beta+1}$, this shows that $\{x, y\}$ is constructible. Thus we have shown: $L(x) \& L(y) \rightarrow L(\{x, y\})$, and so the Pair Set Axiom is true in $\langle L, \in \rangle$.

It remains to show that the Axiom of Choice is satisfied by $\langle L, \in \rangle$. I shall show that for any α, the members of M_α can be written down in a list. If you think of how M was defined, you will see that it is enough to show how the new members of $M_{\alpha+1}$ can be added to a list of those in M_α. To go from M_α to $M_{\alpha+1}$ I took what was already in M_α and added in the sets definable from M_α. Given a list of the members of M_α I just need a listing of the sets definable in M_α to be able to list $M_{\alpha+1}$. Now each member of $\mathrm{Def}(M_\alpha)$ depended only on a formula $\psi(v, v_1, \ldots, v_n)$ and the sets $a_1, \ldots, a_n \in M_\alpha$ which were substituted for the free variables $v_1, \ldots, v_n$. There are a countable number of formulae in our language, so I can certainly write these formulae out in a list. Then, using the list of the members of M_α, I can form a list of all $n + 1$-tuples $\langle \psi, a_1, \ldots, a_n \rangle$ (for all possible values of n). Anything in $\mathrm{Def}(M_\alpha)$ is obtained by using one of these tuples, so the members of $\mathrm{Def}(M_\alpha)$ may be listed in the order that their tuples appear in the list of tuples. This list of the members of $\mathrm{Def}(M_\alpha)$ is now tacked on to the list of M_α to give us a list of $M_{\alpha+1}$.

Using these lists of the members of each M_α, I can now give a list of all the constructible sets as follows. Take constructible sets x and y. Let α and β be, respectively, the least ordinals for which $x \in M_\alpha$ and $y \in M_\beta$. Then x is before y in the list of L if $\alpha < \beta$, or if $\alpha = \beta$ and x is before y in the list of M_α.

Now take any constructible set x, and I shall describe to you a choice function f for x. It is easy to show that if $y \in x$ then y and all the members of y are also constructible, and so occur in our list of L. So, for any non-empty set y from x, let $f(y)$ be the first z in the list of L for which $z \in y$. Then f is indeed a choice function for x.

There is just one more point in establishing that $\langle L, \in \rangle$ satisfies the Axiom of Choice. The function f must be itself in L. This is in fact the case, although it is rather more difficult to show.

This concludes my outline of Gödel's constructible sets—a model of set theory in which the Axiom of Choice is true. The Axiom of Choice is thus consistent with set theory. In fact, the Generalized Continuum Hypothesis is also true in L, and so the GCH is likewise consistent with set theory (including the Axiom of Choice).

In the remainder of this chapter I shall indicate how to show that the Axiom of Choice and the Generalized Continuum Hypothesis are each independent of set theory: neither can be proved from the other axioms of set theory. This is done by finding models of set theory in which each in turn is false. These models were first found in 1963 by Paul Cohen, some sixty years after it was first realized that the Axiom of Choice was an axiom about sets that one needed to have. The fact that it took this long reflects the difficulty in setting up the models, and emphasizes that I shall be presenting only a very rough outline of the method here. I shall in fact show only the result, slightly stronger than the independence of the Generalized Continuum Hypothesis, that there is a model of set theory in which the Continuum Hypothesis is false. I shall not deal with the Axiom of Choice. The method there is similar, just a little more complicated.

You will remember that CH states: $2^{\aleph_0} = \aleph_1$. I can in fact give you a model in which $2^{\aleph_0} = \aleph_2$, or $\aleph_3$, or any one of a lot of other infinite cardinals. My strategy will be this: If we can lay hands on a model of ZF (which includes the Axiom of Choice) in which CH is false, then we have nothing more to do. So I shall suppose that the only models of ZF that we can find are ones in which CH is true. I shall show how to modify one of these, so that CH becomes false.

So let us fix on a countable model $\langle M, \in \rangle$ for ZF, in which CH is true, and I shall assume further that M is a transitive set—that is, if $x \in M$ and $y \in x$, then also $y \in M$. It is no restriction to use such a countable transitive $\in$-model of ZF, for one can show that if ZF is consistent, as we have tacitly assumed, then it has such a model. (The existence of a countable model comes from the Löwenheim–Skolem theorem of Chapter 3.) I ask that M be transitive, for then one can show that what M sees as ordinals are truly ordinals, and in fact are all the ordinals less than some fixed ordinal.

Pictorially I can represent all the sets in the world as fitting inside a cone, up the middle of which runs the 'backbone' of ordinals, in increasing order. The height up the cone that a set falls is a measure of how complicated it is. The degree of complication is given by the level of a set, that is, the ordinal which is at the same height. This notion can be made precise.

Our transitive model M will have all the ordinals up to some point, plus some of the corresponding sets at each level.

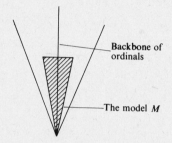

Backbone of ordinals

The model M

THE UNIVERSE OF SETS

In $\langle M, \in \rangle$ there are as few as possible subsets of the natural numbers, for CH is true there. I shall show how to add κ extra subsets, where κ is any fixed cardinal from M, to create a new model $\langle N, \in \rangle$, transitive and with the same ordinals in it. Pictorially, N is a fatter wedge of the same height as M.

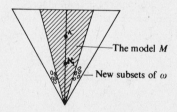

The model M

New subsets of ω

THE NEW MODEL N

I shall describe to you, here in the real world, how to make the construction. In fact, the key point in the proof of the final big theorem that I shall state is the observation that an M-person (that is, someone living in M) could follow too. He will understand, and be able to carry out, some of the instructions that I shall give. Of course, some of what I say will not make sense to him—for instance, when I say that M is countable. M is his whole universe, and so *he* certainly does not think that M is countable. But I shall say no more on this point.

Let us call the new subsets a_η, for $\eta < \kappa$. If N can be constructed so that all the a_η are distinct, then in N we shall have $2^{\aleph_0} \geqslant \kappa$. Of course, if we add these subsets of ω, then there are lots of other sets that we shall have to put in also, if N is to have any hope of being a model for set theory.

In order to be able to talk about N and the new subsets of ω before I have specified them, before I know exactly what they are going to be, I want to have names for all the sets that eventually will be in N. To do this, I shall set up a language whose formation rules give a way of building up names for all the sets that will be in N. Call the language $\mathscr{L}(M)$—it will depend on our original model M. $\mathscr{L}(M)$ is an extension of predicate calculus. It contains:

  (i) constants $\mathbf{a}_\eta$, one for each $\eta < \kappa$
      (to name the new subsets a_η of ω);
 (ii) constants $\mathbf{m}$, one for each $m \in M$
      (to name the elements of M, for we are planning that M will be a subset of N);
(iii) logical signs $\lnot$, &, $\exists$
      (from which the other usual signs can be defined);
 (iv) variables $v, w, \dots$;
  (v) binary predicate symbols ε, $\equiv$
      (to use when we talk about membership and equality);
 (vi) signs $\exists_\alpha$, $\{-:\dots\}_\alpha$, one for each ordinal α in M.
      (The symbol $\exists_\alpha$ is to say there is something in N of level of complexity less than α, with the required property, and $\{-:\dots\}_\alpha$ will be the set of things of level less than α with the property . . . .)

Sentences of $\mathscr{L}(M)$ can be defined as for any predicate calculus, but with a simple extension for $\exists_\alpha$ and $\{-:\dots\}_\alpha$. The model N will have every set in it named by a term from the above list. Some sets may have many such names, but this will not matter.

Using these symbols, we form the terms of $\mathscr{L}(M)$ in the following ways. At the same time, I shall assign to each term an ordinal from M, its *level*, which indicates the largest level of complexity that the set of N eventually to be named by the term may have.

  (i) Each constant is a term. The level of $\mathbf{a}_\eta$ is 1; the level of $\mathbf{m}$ for $m \in M$ is that of the complexity of the set m in M.
 (ii) $\{v:\psi(v)\}_\alpha$ is a term, provided that ψ contains $\exists_\beta$ and $\{-:\dots\}_\beta$ only for $\beta < \alpha$, and does not contain $\exists$. This term has level α. (The restriction on ψ is so that in building up a term of level α we need to know only about things of level lower than α.)

We must now consider what restrictions are needed on the new sets a_η, if we are to have any chance of making N a model of set theory. Suppose we know that $5 \in a_{37}$, or $11 \notin a_3$; what other facts must then hold in N? Considerations of this type lead to the following definition.

Definition. A *condition* p is a finite consistent set of ordered triples $\langle n, \eta, i \rangle$

where $n < \omega$, $\eta < \kappa$ and $i = 0, 1$. 'Consistent' means that if $\langle n, \eta, 0 \rangle \in p$, then $\langle n, \eta, 1 \rangle \notin p$, and vice versa.

Each condition p can be thought of as encoding a small amount of information about the model N. If $\langle n, \eta, 0 \rangle \in p$, then this means that $n \in a_\eta$: if $\langle n, \eta, 1 \rangle \in p$ then $n \notin a_\eta$. If q is another condition and $p \subseteq q$, then q gives more information than p. I shall say that q extends p.

The information encoded in a condition p may cause more complicated things to hold in N. We can talk about N by using the language $\mathscr{L}(M)$, so let us define a relation between conditions p and sentences ψ of $\mathscr{L}(M)$, which ought to hold just when the information encoded in p causes what ψ says about N to come true. This relation is written $p \Vdash \psi$, and is read 'p forces ψ'.

(i) $p \Vdash \mathbf{n} \, \varepsilon \, \mathbf{a}_\eta$ if and only if $\langle n, \eta, 0 \rangle \in p$. (I said above that I wanted $n \in a_\eta$ only when $\langle n, \eta, 0 \rangle \in p$.)

(ii) $p \Vdash \mathbf{l} \, \varepsilon \, \mathbf{m}$ if and only if $l \in m$, m in M,

(iii) $p \Vdash \mathbf{l} \equiv \mathbf{m}$ if and only if $l = m$, for l, m in M. (These two clauses occur because we want $M \subseteq N$, and so we do not want to effect membership or equality between members of M.)

(iv) Here come a number of other clauses, dealing with the remaining basic formulae in $\mathscr{L}(M)$. They are a little complicated, so I shall not list them.

(v) $p \Vdash (\psi \, \& \, \theta)$ if and only if $p \Vdash \psi$ and also $p \Vdash \theta$.

(vi) $p \Vdash \exists v(\psi(v))$ if and only if $p \Vdash \psi(t)$, for some term t. (In other words, p causes $\exists v(\psi(v))$ to hold just when p causes some set to have the property ψ.)

(vii) $p \Vdash \exists_\alpha v(\psi(v))$ if and only if $p \Vdash \psi(t)$ for some term t of level less than α. (As for (vi), except that the set in N with the property ψ must be of level less than α.)

(viii) $p \Vdash \neg \psi$ if and only if for all q extending p it is not the case that $q \Vdash \psi$. (This is the most interesting case of all. We don't say $p \Vdash \neg \psi$ if and only if it is not the case that $p \Vdash \psi$, for maybe p fails to force ψ because p does not encode enough information to settle ψ. However, by adding more triples to p, perhaps we could then force ψ. We want p to force $\neg \psi$ only when, no matter how much additional information we have, ψ is still not forced.)

As an example of how this definition works, let me show you that $p \Vdash \neg(\mathbf{n} \, \varepsilon \, \mathbf{a}_\eta)$ if and only if $\langle n, \eta, 1 \rangle \in p$, as you would expect from the earlier discussion. Now if $\langle n, \eta, 1 \rangle \in p$, no q extending p can have $\langle n, \eta, 0 \rangle \in q$ and so no q extending p can force $\mathbf{n} \, \varepsilon \, \mathbf{a}_\eta$. Hence $p \Vdash \neg(\mathbf{n} \, \varepsilon \, \mathbf{a}_\eta)$.

On the other hand, suppose $\langle n, \eta, 1 \rangle \notin p$. Consider $q = p \cup \{\langle n, \eta, 0 \rangle\}$. q is consistent and finite (since p is), and so is a condition. Since $\langle n, \eta, 0 \rangle \in q$, we have $q \Vdash \mathbf{n} \, \varepsilon \, \mathbf{a}_\eta$, and so by clause (viii), $p \Vdash \neg(\mathbf{n} \, \varepsilon \, \mathbf{a}_\eta)$.

I want now to consider finding a set G of conditions such that the information encoded in the conditions in G is sufficient to determine in advance all the things that will be true in N although we do not know yet what N is to be. That is, I want:

(a) For each sentence ψ in $\mathcal{L}(M)$, there is $p \in G$ such that $p \Vdash \psi$ or $p \Vdash \neg\psi$.

   Clearly, G must do this consistently, that is

(b) For no p, q from G and sentence ψ in $\mathcal{L}(M)$ do we have both $p \Vdash \psi$ and $q \Vdash \neg\psi$.

A set G of conditions is called *generic* if it has the properties (a) and (b). One can show that in fact such a generic set does indeed exist. (This is the only point at which one uses the fact that M is countable.)

Now fix on a generic set G. I shall use it to give an interpretation, as a particular set, to each of the terms in the language $\mathcal{L}(M)$, and then, finally, our set N will be just the set of all these interpretations. (Thus t has been a name for its interpretation.) So for each term t, I define its interpretation $I(t)$ as follows:

(i) $I(\mathbf{a}_\eta) = a_\eta = \{n \in \omega : \exists p \in G(\langle n, \eta, 0 \rangle \in p)\}$. (Thus a_η is the set of those numbers n for which some p in G encoded that n should belong to a_η.)

(ii) $I(\mathbf{m}) = m$, for each m in M. (For $\mathbf{m}$ was always thought of as a name for m.)

(iii) If we already know $I(t)$ for each term t of level less than α, then $I(\{v : \psi(v)\}_\alpha)$
   $= \{I(t) : t \text{ is a term of level less than } \alpha, \text{ and } \exists p \in G(p \Vdash \psi(t))\}$.

(So $\{v : \psi(v)\}_\alpha$ is interpreted as the set of those things of level less than α which some p in G forces (their name) to have the property ψ.) Now let N be the collection of all the interpretations. If we now interpret ε and $\equiv$ as truly $\in$ and $=$, we can prove that what we have been working for actually does happen.

THEOREM (Cohen's Truth Lemma): A sentence ψ of $\mathcal{L}(M)$ is true in N under this interpretation if and only if there is some $p \in G$ such that $p \Vdash \psi$.

Consider now the structure $\langle N, \in \rangle$. Several marvellous things happen. I shall sum them up in the following theorem.

THEOREM (Cohen): (i) $\langle N, \in \rangle$ is a model of ZF. (ii) The ordinal numbers and cardinal numbers of $\langle N, \in \rangle$ are the same as those of $\langle M, \in \rangle$. (iii) $\langle N, \in \rangle \models 2^{\aleph_0} \geqslant \kappa$.

As I hinted earlier, this theorem holds because in fact an M-person can understand and write down the definition of forcing. The Truth Lemma above thus shows that questions about N can be rephrased as questions concerning M. Since M is a model of set theory, we know the answers there.

Part (iii) is not as hard as the other two parts. It is not very difficult to show that in fact the a_η are all distinct, that is, if $\eta \neq \zeta$ then $a_\eta \neq a_\zeta$. Thus we have κ different subsets of ω, and so $2^{\aleph_0}$ must be at least κ. If we take $\kappa = \aleph_2$ (or indeed, $\aleph_n$ for any finite n) then it turns out that $\langle N, \in \rangle \models 2^{\aleph_0} = \kappa$. So we can have a model of set theory in which $2^{\aleph_0} = \aleph_2$, say. Any such model suffices to establish the independence of the Continuum Hypothesis from the set theory ZF.

Some Suggestions for Further Reading

ANY thorough treatment of the topics covered in this book involves quite hard mathematics and none of the books listed below is particularly easy to read (which is one of the reasons we wrote our book).

	COVERS PARTS OF CHAPTERS
E. J. LEMMON Beginning logic. (Nelson 1965)	2
G. T. KNEEBONE Mathematical logic and the foundations of mathematics. (van Nostrand 1963)	1, 2, 4, 5, 6
R. R. STOLL Set theory and logic. (W. H. Freeman 1963)	2, 5
A. MARGARIS First order mathematical logic. (Blaisdell 1967)	2
J. W. ROBBIN Mathematical logic, a first course. (Benjamin 1969)	2, 5
A. I. MAL'CEV Algorithms and recursive functions. (Wolters-Noordhoff 1970)	4
R. C. LYNDON Notes on logic. (van Nostrand 1966)	2, 5
J. R. SHOENFIELD Mathematical logic. (Addison-Wesley 1967)	1, 6
E. MENDELSON Introduction to mathematical logic. (van Nostrand 1964)	2, 5
P. J. COHEN Set theory and the Continuum Hypothesis. (Benjamin 1966)	5, 6

Index

PRINTED IN GREAT BRITAIN
BY J. W. ARROWSMITH LTD.
BRISTOL, ENGLAND